Engineering Education at Penn State

A Century in the
Land-Grant Tradition

Michael Bezilla

The Pennsylvania State University Press
University Park and London

Photographs contained in this volume are courtesy of Purdue University, Penn State Collection, College of Engineering, and *La Vie* (as indicated in captions), and are reproduced by permission.

Library of Congress Cataloging in Publication Data
Bezilla, Michael.
   Engineering education at Penn State.
   Includes bibliographical references and index.
   1. Pennsylvania State University. College of Engineering. I. Title.
T171.P54B49   620'.007'1174853   81-47170
ISBN 0-271-00287-5   AACR2

Copyright © 1981 The Pennsylvania State University

All rights reserved

Designed by Dolly Carr

Printed in the United States of America

Contents

Preface		vii
1	**The Formative Years: 1855–95**	1
	Evan Pugh and Lost Opportunities	2
	The Calder Era	9
	Atherton and Engineering	13
	Formation of the School of Engineering	32
2	**Growth and Maturation: 1895–1915**	37
	Coping with the Demand for Technical Education	37
	President Sparks and Dean Jackson	54
	Beginnings of Research and Extension Education	64
3	**Consolidation and Retrenchment: 1915–29**	77
	New Leadership for the Crisis Years	78
	An Attempt for Renewed Growth	94
	Reaffirmation of the Importance of Engineering	107
4	**Depression and War: 1929–45**	115
	Progress Amid Adversity	118
	Harry Hammond: Entering a New Epoch	127
	Engineering and Another War	138
5	**New Directions: 1945–56**	147
	Foundations for the Future	149
	Eric Walker and an Era of Modernization	160
	Curricular and Administrative Changes	169

6	**Adapting to Changing Demands: 1956–81**	181
	Emphasis on Research	183
	Restructuring the College	190
	The Palladino Years	200
	Retrospect	208

Notes 213

Appendixes

A.	Deans of the College of Engineering	219
B.	Engineering Department Heads	221
C.	College of Engineering Enrollment since 1895	225
D.	Total Enrollment at Penn State since 1895	227

Index 229

Preface

This book originated as a commissioned project of the Penn State Engineering Society, the alumni body of the College of Engineering. While focusing principally on the College, I have endeavored to place the development of engineering at The Pennsylvania State University in the context of state and national events, thereby producing a case study that would be of value to historians and educators as well as to individuals having a personal connection with the College. With the exception of a few recent investigations, historians have ignored the evolution of engineering education, despite the pivotal role it has played and continues to play in a society whose destiny has been so profoundly shaped by technology. Furthermore, I had hardly commenced work on the project when I discovered that engineering education at Penn State has not received attention commensurate with the influence it exerted on the character of the University, which many persons mistakenly picture as mainly an agricultural institution from its founding in 1855 at least until World War II. In reality, engineering was an extremely important component of land-grant education at Penn State and most other schools in the land-grant system.

I have not hesitated to lay bare the weaknesses of the College of Engineering and its leaders and have been allowed the utmost freedom to do so by the project's sponsors. If I have been sparing in my criticisms, it is only because a careful examination of the historical record reveals so little that can be faulted during the early years of the College. Events of the last two or three decades may perhaps be another matter, but one lacks the perspective to attempt a critical analysis of this period. Only a historian belonging to a later generation can adequately chronicle these years, and for that reason I have not treated the recent era in the same detail as earlier periods.

I am indebted to Nunzio J. Palladino, former Dean of the College of Engineering of The Pennsylvania State University, and to Walter G. Braun, Associate Dean Emeritus, for supplying a wide range of logistical support and for making available the records of the College. In addition, Deans Palladino *and Braun and Leon J. Stout, the University's archivist, read the manuscript and provided constructive criticism. Gifford H. Albright, Joseph C. Conway, O. Allan Gianniny, Jr., William H. Gotolski, C. B. Holt, Jr., Benjamin Niebel, Eric A. Walker, and Merritt A. Williamson read all or parts of the manuscript and made suggestions for its improvement. Mr. Stout was helpful in putting at my disposal the resources of Pattee Library's Penn State Room, as were Cynthia J. Ahmann and other members of the Penn State Room staff. I also wish to thank the many alumni and present and former members of the faculty and staff of the College of Engineering who shared their knowledge of the College's history and personalities with me either through personal interviews or written correspondence and who helped me in other ways too numerous to mention. Not to be overlooked, either, are the contributions of Sharon Dailey and Sue Doland, who typed several versions of the manuscript. A special note of appreciation must go to Richard E. Grubb, Senior Vice-President for Administration of The Pennsylvania State University, for permitting me to put aside other responsibilities in order to see this project through to its completion.*

1 The Formative Years: 1855–95

The history of the College of Engineering of The Pennsylvania State University begins long before the institution held its first formal classes in engineering. In 1854, the Pennsylvania state legislature granted a charter to The Farmers' High School, whose purpose, as expressed in its first catalog (1859), was to "adopt a system of instruction which shall embrace, to the fullest extent possible, those departments of all sciences which have a practical or theoretical bearing upon agriculture." For a decade or more prior to the granting of this charter, sentiment had been building among Pennsylvania farmers and related interests for an institution that offered advanced, practical instruction in scientific agriculture—a subject absent from the courses of study then available at most colleges and universities. The catalyst of this movement was the Pennsylvania State Agricultural Society, organized at Harrisburg in 1851. Bringing systematic pressure to bear on the Commonwealth's political representatives, the Society encountered little opposition to the idea of an agricultural school and easily won official sanction for it. Although the legislature issued a slightly revised charter to the School on February 22, 1855 (a date Penn State observes as its founding), the new document reiterated the call for students "to be instructed and taught all things necessary to be known as a farmer." By 1858, the board of trustees of The Farmers' High School had selected a site of some 200 acres in rural Centre County, near the geographic center of the state. A legislative appropriation of $25,000 (with an additional $25,000 pledged), coupled with other large sums raised through public subscription and private donation, provided enough money to begin construction of the main building. On February 16, 1859, with four faculty members in attendance, the school admitted its first class of 69 students.

Evan Pugh and Lost Opportunities

To head the institution, the trustees selected Evan Pugh, a 31-year-old native of Chester County who had spent the last few years studying in Europe. While his father had been a farmer-blacksmith and he himself had worked on a farm as a youth, Pugh was not an agriculturalist in the strict sense. He had devoted much of his time to exploring methods of applying science to agriculture and had received a Ph.D. in agricultural chemistry from Germany's University of Goettingen in 1854. He then moved to England, where his investigations in the fields of chemistry and agriculture won international acclaim and brought him to the attention of the Farm School's trustees. While Pugh welcomed the opportunity to develop the school into a leading center of agricultural teaching and research, he nevertheless envisioned a broader objective for the institution, and in that vision lie the roots of engineering education at Penn State. President Pugh wished to see the school expand its educational offerings beyond the narrow confines of agriculture to a full range of subjects that would be equally useful to the citizens of a maturing nation. "The fundamental idea is to associate a high degree of intelligence with the practice of agriculture and the industrial arts," he wrote in 1862, "and to seek to make use of this intelligence in developing the agricultural and industrial resources of the country, and protecting its interests."[1] Following the custom of the day, Pugh defined "industrial arts" to mean both those subjects that required a high degree of theoretical understanding, such as surveying, and those that demanded manual dexterity, such as carpentry.

Few American colleges were as yet prepared to offer their students anything other than the kind of classical learning, with its emphasis on ancient languages, philosophy, rhetoric, and similar subjects, that had formed the core of the higher education curriculum for over two hundred years. This type of education was growing increasingly inadequate for a nation that expected to tame a hostile natural environment of immense proportions and to ensure material as well as political democracy. College educators were slow to yield to the changing needs of society, however. As late as 1860, fewer than a dozen institutions offered baccalaureate programs in engineering and had graduated a combined total of only two hundred or so students. These institutions adopted varying approaches to engineering education. A few—Rensselaer Polytechnic Institute and Norwich University, to name the most prominent—specialized in training engineers, while others—the University of Michigan, for example—included engineering as just one of several technical and non-technical courses of study available to the prospective student. Still other institutions established semi-autonomous academic divisions to administer nearly all their engineering and scientific studies. This dichotomy reflected the widespread belief that the utilitarian nature of the technical curriculum rendered it inferior to the classical curriculum, which had as its objective the mental and moral improvement of the student. The view that integrating engineering and other technical subjects with the classics might somehow subvert the purpose of a college education was particularly prevalent at older, more tradition-bound institutions such as Harvard, Yale, and Dartmouth, all of which had created separate "scientific schools" in the years just before the Civil War. Whatever the arrangement, almost all instruction was limited to the category of civil engineering, with canal, railroad, and bridge construction attracting the interest of most graduates.[2]

Even so, the vast majority of civil engineers still learned their profession through apprenticeships and other forms of on-the-job experience. Evan Pugh realized that this informal system of education could no longer cope with the intensifying demand for persons conversant with the mechanic arts, as engineering was frequently termed. Colleges must be equipped to provide the technological training that America required to maintain its economic growth and geographic expansion. Would not The Farmers' High School be performing a vital service to the Commonwealth and to the nation, Pugh reasoned, by educating students in agricultural and industrial pursuits? Pugh's was not the only voice to be raised in behalf of adding subjects having practical worth to the college curriculum. In July 1862, Congress passed the Morrill Land Grant Act, a statute sponsored by Vermont's Representative Justin S. Morrill and heartily endorsed by Pugh and other reform-minded educators. The act, which was to have profound implications for engineering education at Penn State and throughout the United States, read in part:

> The leading object shall be, without excluding other scientific and classical studies, and including military tactics, to teach such branches of learning as are related to agriculture and the mechanic arts, in such a manner as the legislature of a state may prescribe, in order to promote the liberal and practical education of the industrial classes in the several pursuits and professions of life.

4   Formative Years, 1855–95

Evan Pugh (Penn State Collection)

Each state, upon acceptance of the provisions of the act, was to be granted 30,000 acres of land in the national domain for each senator and representative it had in Congress. The state was to sell the land, permanently retain the proceeds, and use the income generated therefrom to endow one or more colleges having four-year degree curriculums. In return, these colleges were to satisfy provisions of the act by making available the appropriate courses of study.

Here was the opportunity Pugh had been awaiting. Relying on its own meager resources, The Farmers' High School would be unable to enlarge its curriculum, at least in the foreseeable future. Drawing on the land-grant endowment, expansion beyond the confines of agriculture appeared a certainty. But first Pugh had to convince the Pennsylvania legislature to designate the school as the beneficiary of the Commonwealth's land-grant funds. He had already taken steps to upgrade the image of the institution by changing its title to The Agricultural College of Pennsylvania. The new name more accurately reflected the fact that the "high school" was and always had been a college with a four-year course of studies leading to a bachelor's degree. To be sure, the Agricultural College was not without rivals for the land-grant endowment. Chief among them was the Polytechnic College of the State of Pennsylvania. All but forgotten today, the Polytechnic College was the brainchild of Dr. Alfred Kennedy, a physician and chemist who founded the school in Philadelphia in 1853 to train "professional miners, engineers, and directors of farms and factories." Other colleges in the Commonwealth, including Allegheny, Lafayette, Dickinson, and the Western University of Pennsylvania (later the University of Pittsburgh), offered sporadic instruction in civil engineering but merely as a component of a "general scientific" course that also contained large doses of the natural sciences. Only the Polytechnic College offered a baccalaureate program designed exclusively for students who wished to become engineers—and not necessarily civil engineers, either. The Polytechnic College was the first institution in the United States to establish degree curriculums in mechanical engineering (1854) and mining engineering (1857). Kennedy patterned his school after the great polytechnic institutes of Europe and hoped that, like these models, the Polytechnic College would eventually gain state financial support. Its claim to a share of Pennsylvania's land-grant money was without question more justifiable than those of any other school in the Commonwealth with the exception of the Agricultural College, since Kennedy had already initiated instruction in the mechanic arts and thus partially satisfied the provisions of the Morrill Act. His efforts to launch an agricultural curriculum consistently ended in failure, however, owing to an absence of facilities for a model farm and a lack of interest in farming among the city-bred boys of Philadelphia. By contrast, the Agricultural College, which had a flourishing course in agriculture, was not likely to meet insurmountable obstacles in its attempt to begin training in the mechanic arts.[3]

In the spring of 1863, Pennsylvania accepted the stipulations of the Morrill Act and named the Agricultural College as the recipient of all land-grant revenues. Instruction in the mechanic arts could not begin immediately, however. Because of the long period of time required to sell all the land allotted to the Commonwealth, the College could not expect to receive a substantial payment from its endowment for several years. A far more crippling setback occurred in April 1864, when President Pugh died of typhoid fever. His passing dealt such a devastating blow to engineering that over two decades were to pass before his vision of the College as a center for industrial education would begin to be realized.

Elected to succeed Pugh in June 1864 was William H. Allen, former president of Philadelphia's Girard College and one-time Professor of Chemistry at Dickinson College. Allen exhibited little interest in continuing his predecessor's struggle to broaden the curriculum of the College. Instead, he devoted most of his attention to the political arena, where he fought to preserve the College's designation as the Commonwealth's only land-grant institution. Other Pennsylvania colleges, envious of Penn State's apparently privileged status, were campaigning to be named co-beneficiaries of the land-grant monies. No real champion of the mechanic arts appeared until John Fraser accepted an appointment as Professor of Mathematics in July 1865. Fraser, a Scottish immigrant, had served for several years as a mathematics professor at Jefferson College, from which post he resigned in 1862 to join the Union Army. Soon after coming to the Agricultural College, Fraser recognized that it was not complying with the mandate of the Morrill Act. True, the trustees had added military tactics to the curriculum (Fraser himself was instructor), but little had been accomplished toward providing instruction in engineering. Professor Fraser became such a vocal advocate of curricular reform that upon Allen's resignation in September 1866, the board of trustees elected him to fill the vacancy.

Fraser did not stand alone in his desire to see a resurrection of Pugh's concept of industrial education. In May 1866, a six-member faculty-trustee committee chaired by Fraser drew up a plan for reorganizing the curriculum. As president, the determined Scot pushed hard to have the trustees adopt this plan. By the spring of 1867, he had succeeded, as the board finally agreed to supplement the lone course in agriculture with several new courses of study. The four courses* to be added in the 1868–69 academic year were general science, literature, mechanical and civil engineering, and metallurgy, mineralogy, and mining. Each was to be a complete four-year curriculum leading to a Bachelor of Science degree.

The new president discovered to his dismay that approval of these changes in principle did not necessarily mean that they would be implemented according to schedule. Many of the trustees were agricultural men and resented any attempt to "dilute" the original goals of the College. Others—the majority—sympathized with Fraser's intentions but believed that the College was not yet in a position to undertake significant reforms. Student enrollment had shown a marked decline in recent years, falling from 146 degree students for 1864–65 to a mere 82 for the 1866–67 sessions. As the number of students fell, so too did the College's income. Steadily shrinking tuition receipts forced the trustees to issue $80,000 worth of bonds in 1866. The legislature offered no assistance, notwithstanding the fact that many of its members criticized the way College affairs were being handled. Even the

---

*Until otherwise indicated, the term "course" refers to a complete degree curriculum and not to a series of instruction periods related to one specific subject.

endowment from the Morrill Act, just now beginning to trickle in, was much smaller than expected because of the depressed market for land. President Fraser countered that the decrease in enrollment indicated that potential students were finding the purely agricultural curriculum of the College unattractive and were going elsewhere for their education. One could not logically expect a return to financial soundness, he maintained, until students came to the College in larger numbers, and that would not occur unless the school enlarged its course offerings, particularly in the area of engineering. The trustees had their way in the end. Reality dictated a cautious approach to curricular reform. The catalog for 1868–69 made no mention of the proposed mining course, although it did list one in mechanical and civil engineering. That course might just as well have been deleted, too, since the trustees had hired no faculty to provide instruction. A defeated President Fraser, noting the existence of certain "irreconcilable differences with the board of trustees as to scope and policies of the institution," submitted his resignation on March 4, 1868.[4] (He subsequently became Chancellor of the University of Kansas and still later Professor of Mathematics at the Western University of Pennsylvania.) With Fraser's departure went what little hope had remained after Pugh's death that the College still might be able to establish a program of engineering education in the near future.

John Fraser's successor brought to his post a unique concept of curricular reform that at least was in keeping with the desire of the people of the Commonwealth, their elected representatives, and many of the trustees and faculty to broaden the course offerings of the College. Thomas H. Burrowes, Penn State's fourth president, was a former legislator and farmer who had earned high regard as state Superintendent of Schools and editor of the *Pennsylvania School Journal*. He proposed that the curriculum of the College be divided into three courses of study, with subject areas dependent on the number of years a student wished to pursue studies. At the end of their third year of instruction, which consisted primarily of agricultural topics or the "Agricultural Course," students were awarded a Bachelor of Scientific Agriculture degree. Students choosing to go on for an additional year would take the "Scientific Course," leading to the degree of Bachelor of Science. This course, as described in the 1868–69 catalog, was "for civil engineers, general mechanics, etc.," as well as for those interested in pure science. Students could also elect a fifth year of studies, thus entering the "Literary Course," which contained the kind of classical education students received at most other colleges and culminated with the Bachelor of Arts degree.

President Burrowes never had the opportunity to see how well this peculiar scheme would function. He died in February 1871 from exhaustion and exposure after accompanying students on a mid-winter outing. Had he lived longer, however, there is little evidence to intimate that the seeds of engineering education would have at last found a fertile field under his stewardship. The "Scientific Course" was something of a catch-all, and except for land surveying and analytical mechanics, included little formal instruction in engineering. No faculty member was hired exclusively for the purpose of teaching engineering subjects. What scant instruction that did exist was in charge of a Professor of Mathematics and Civil Engineering. Isaac Thomas held that post until succeeded in 1870 by Israel Heilman, who was in turn replaced by Reverend Francis Robinson the following year. None of these men had much academic or practical experience in engineering.

Evidence that the Agricultural College possessed minimal attractions for students who had no interest in farming could be seen in the continual decline in enrollment. The start of the 1870–71 school year found just 59 degree students in attendance, all but seven of whom were in their first or second years of study. Agriculture itself was insufficient to support a college, as the experience of many other land-grant institutions was proving. Practical farmers, ignorant of the value of scientific agriculture, saw little reason to send their sons to school to learn "book-farming." Substantial additions to the curriculums of these institutions were needed, not only to bring the colleges into conformity with the Morrill Act, but to ensure their very survival. Unfortunately, the fact that enrollments were so low made the task of financing curricular reform a most formidable one, particularly in a state such as Pennsylvania, where the legislature expressed a decided indifference toward land-grant institutions. The legislators had not yet adopted the opinion that by reason of its selection as Pennsylvania's land-grant school the Agricultural College was a child of the Commonwealth and therefore should be supported by regular state appropriations. Consequently, while demands that the College diversify its educational offerings emerged from Harrisburg with greater and greater frequency, the money to help implement this diversification did not. Compounding the College's dilemma was its own ineffective leadership. Since Pugh's death, only President John Fraser had expressed a willingness to make the necessary curricular changes that would transform the institution into the kind of college that Pennsylvania, the manufacturing, mining, and transportation hub of the nation, could depend upon to preserve its economic preeminence. Now, with the death of President Burrowes, the trustees had the opportunity to select a strong, farsighted leader who would at last take on the unfinished work of Evan Pugh.

The Calder Era

On March 10, 1871, the trustees elected James Calder as the fifth president of the Agricultural College of Pennsylvania. Calder, a Harrisburg native and a Methodist minister, came to the Agricultural College after a tenure as president of Hillsdale College in Michigan. He quickly eliminated Burrowes' organization of three-, four-, and five-year courses and returned to the traditional four-year program. To his credit, Calder recognized the weaknesses of a purely agricultural course of instruction. Under his guidance, the College for the first time offered baccalaureate courses in non-agricultural subjects, and in 1874 adopted the name Pennsylvania State College to underscore its newly broadened curriculum. Nevertheless, the courses were not in harmony with the primary thrust of the Morrill Act. Besides the standard agricultural course, the curriculum included a scientific course and a classical course. Neither of the new additions differed in many respects from the traditional studies found at other colleges and universities. Virtually no provision was made for the mechanic arts, except as certain elements, mainly surveying and drafting, related to topics studied in the agricultural and scientific courses. As in the past, the Professor of Mathematics and Civil Engineering had the responsibility of teaching these subjects. John F. Downey succeeded Professor Robinson to the Mathematics and Civil Engineering chair in 1874 and continued in that post until 1880. Downey confessed that the "Civil Engineering" portion of his title was primarily a courtesy, since he taught only a few aspects of that subject and then "not with the view of making engineers, but with the view of supplying information which every practical man should possess."[5] Such a curriculum in no way justified the College catalog's description of it as "a thorough course of instruction in mechanical and civil engineering, in the interest of the mechanical arts." And while the catalog might also proclaim that students had ample opportunity to observe "applications of knowledge" at such nearby locations as the Oak Hall woolen factory, the Bellefonte water and gas works, and the coal mines near Snow Shoe, they received precious little instruction in any kind of knowledge they could apply. Downey's classes were confined to the basement of the Main Building (the original Old Main) and laboratories and practicums were non-existent. In sum, President Calder had succeeded in broadening the educational base of the institution, but he failed to expand in those directions most desired by many of the College's friends and supporters.

Surveying practicum, 1880. The woman in the second row is Esther Hunter, the first female to graduate from the scientific curriculum, for which the practicum was a requirement. (Penn State Collection)

The Calder administration has often been described as representing the darkest hour in the entire history of Penn State, a time of weak leadership and drift even further from the ideals of land-grant education. Certainly this was true in regard to engineering. Other schools were meeting the challenge of the industrial age by establishing engineering courses in a variety of fields. In Pennsylvania, Lehigh University, founded by railroad magnate Asa Packer in 1866, offered baccalaureate programs in civil, mechanical, and mining and metallurgical engineering, while the Western University of Pennsylvania in 1871 initiated a four-year curriculum in civil engineering. The number of colleges and universities in the United States that possessed engineering departments rose to seventy by 1872, with well over half being supported by land-grant endowments.[6] The Pennsylvania State College, despite its obligations under the Morrill Act, chose instead to imitate the classical institutions, whose lack of offerings in practical subjects had been the very reason for establishing the College.

As the Calder years wore on, The Pennsylvania State College's role as a serious and vital institution of higher learning came under attack in the legislature and throughout the Commonwealth. Popular confidence in the College plummeted. Only four students graduated in 1875, seven in 1876, and three in 1877. As late as 1880, the College was, in the words of Penn State historian E.W. Runkle, "still floating in the stagnant backwaters of pseudo-classical and literary retreats." By that time, dissatisfaction among trustees, faculty, and the legislature with Calder's handling of affairs had grown so intense that late in 1879 the president tendered his resignation.

Joseph Shortlidge, Calder's successor, was a secondary school principal from Delaware County and not well-versed in matters of higher education. Still, with his assumption to the presidency, many of the trustees and faculty expected substantial reforms to be made in the curriculum. At its meeting of January 27, 1881, the trustees appointed a five-member joint trustee-faculty committee headed by trustee J.P. Wickersham to reorganize the College's courses of study. The committee was to recommend the addition of new courses and the elimination of existing ones in order to better equip the College to satisfy the needs of the nation's most industrialized state. The time for such reorganization was long overdue. As one disgruntled professor exclaimed, "After twenty-two years of experimenting, we are today the laughing stock of the state—as an industrial college, we are a failure."[7] The committee was slow in commencing its work, however, and the deliberateness with which it pursued its inquiry aroused the ire of many of the reform-minded members of the teaching staff.

Shortly after the Wickersham committee was formed, President Shortlidge appointed a three-man faculty committee to conduct a similar investigation of the curriculum and make recommendations for improvement. Shortlidge made the appointments with considerable reluctance and only at the rigid insistence of a large segment of the faculty, who were dissatisfied with the slowness of the Wickersham investigation. Jealous of his prerogatives, Shortlidge feared that the faculty would use their own panel as a tool to enlarge their power at the expense of the president's. To guard against such encroachment, he selected three professors to serve on the committee whom he believed to be at best lukewarm toward any significant reorganization. About Josiah Jackson (Professor of Mathematics since Downey departed in 1880) and J.M. Campbell (Professor of Latin), Shortlidge judged correctly. His third selection, I. Thornton Osmond, Professor of Physics, expressed enthusiasm for change, arguing that the College "had nothing whatever in mechanic arts and engineering" and therefore was not in compliance with the Morrill Act nor with the wishes of the citizens of the Commonwealth. Osmond was chagrined to find that neither of his colleagues had much interest in working for curricular reform. Frustrated by their stance and disappointed at the delays of the Wickersham committee, he marshaled the support of sympathetic faculty members and a few trustees and launched an unofficial reorganization study—"entirely irregular and without authorization," Osmond later admitted.[8]

Osmond and his colleagues prepared to report on their plan for reorganization when the board of trustees convened in special session on April 8, 1881, to hear the findings of the Wickersham group. Soon after the meeting began, President Shortlidge, piqued that these findings dared to criticize his own brief administration, submitted his resignation. The trustees accepted it at once, and no doubt with considerable relief. A quarrelsome and obstinate figure, Shortlidge had no talent for a post demanding tact and the ability to compromise. The very existence of Osmond's committee showed how strained relations between the faculty and the president had become. Professor Osmond eventually presented his report to Professor of English Literature and Greek James Y. McKee, whom the trustees had appointed acting president, and who just happened to be a member of Osmond's own committee! McKee recommended adoption of the Osmond plan, since it and the Wickersham findings were virtually identical in the area of curricular reform. The trustees concurred, giving their formal blessing to a reorganization of the curriculum at their regular meeting in June 1881.

The new curriculum, to take effect with the forthcoming academic year, established six courses of study. Two of these were classified as "general" (the scientific and the classical, holdovers from Calder's day) and four were "technical" (agricultural, natural history, chemistry and physics, and civil engineering). To head the Department of Civil Engineering, the trustees appointed Louis A. Barnard, a Naval Academy graduate who had acquired much civil engineering experience with the Navy and later in a civilian capacity with the federal government. He was given an annual salary of $1200. The trustees also approved a two-year course in mechanic arts, which at this infant stage consisted primarily of classes in mechanical drawing. John H. McCormick, a recent graduate of the Ohio State University, was secured as instructor. He departed after a single term to accept a higher paying position elsewhere, so for the next two years, Professors Barnard and Josiah Jackson shared the duties of teaching mechanical drawing.

The trustees, heartened by the reforms and eager to publicize the expanded curriculum, asked the General Assembly to investigate affairs at the College. Only a short time before the trustees would have feared such a probe, knowing full well Harrisburg's growing dissatisfaction with the direction in which Calder had taken the school. The trustees now calculated that an investigation would vindicate the College and put an official seal of approval on the curricular reorganization, while at the same time boosting the institution's popular esteem and attracting more students. It was a shrewd maneuver, one that would pay dividends in years to come not only for engineering education but for other fields as well. The investigating committee, lead by Senator C. T. Alexander of Centre County, completed its work by the end of 1881. In a report submitted to the legislature in February 1882, the committee acknowledged that The Pennsylvania State College had encountered "widespread distrust, if not hostility," in past years, but concluded that these feelings "grew out of a condition of things which no longer exist." Moreover, the report labeled the $30,000 annual income the College was receiving under the Morrill Act "a mere pittance" and urged the legislature to make periodic and generous appropriations to Penn State.[9] By accepting the findings of the committee's report, the legislature in effect recognized its obligation to support the College financially. In consequence, although Penn State would often be hard-pressed to obtain adequate funds in the years to come, it would never again experience hardships of the severity that it had heretofore endured.

In testifying before the investigating committee in 1881, College trustee S.W. Starkweather spoke optimistically of the institution's future and asserted that the only remaining obstacle to carrying out the full intent of the Morrill Act was "the want of a president thoroughly acquainted with farming and the mechanical arts . . . a president who really felt thoroughly interested in making farming and mechanics a source of importance to the young rising youth of the state."[10] Not long after Starkweather made these remarks, the era of presidents who either had no interest in industrial education, or did not hold office long enough to imprint whatever interest they did have, came to a close. With the appointment of George W. Atherton to the presidency during the summer of 1882, The Pennsylvania State College entered a well-deserved period of administrative stability and internal growth. Under the steady hand of President Atherton, engineering education took firm root at the College and eventually became the source of much academic distinction and national renown.

Atherton and Engineering

The 45-year-old Atherton was a Massachusetts native who for the past fourteen years had been teaching political science at New Jersey's

George W. Atherton (Penn State Collection)

Rutgers College. Long before assuming his duties at Penn State, Atherton had displayed a deep interest in and enthusiasm for land-grant education. As president of a land-grant institution, Atherton was to be extremely influential in forthcoming years in promoting a broad spectrum of higher education at land-grant colleges and universities across the country. Closer to home, Atherton hoped to see The Pennsylvania State College become the capstone of the Commonwealth's entire public school system. As a state-supported institution, the College should and one day would, he believed, offer diverse educational opportunities of the highest quality to the citizens of Pennsylvania.

The new president gave first priority during the early years of his administration to enlarging the College's undergraduate engineering program. The Department of Civil Engineering had been created the year prior to his taking office. Not long after Atherton arrived, Professor Barnard greeted him with a list of $3000 worth of laboratory and other practical materials his department must have in order to provide satisfactory instruction. The president agreed wholeheartedly with the request. "The Department of Civil Engineering has been established only a little more than a year, but is proving, as was anticipated, to be one of the most attractive in the College to active and ambitious young men," he observed in his annual report for 1882. "It is of very great importance to the permanent reputation and interests of the institution that it be furnished with every appliance needed to make its work effective." Barnard soon received all the equipment he had asked for. Most of this gear was to be used for the practicums, or laboratory sessions, where students had the opportunity to use the abstract knowledge acquired in the classroom to solve practical problems. All students, regardless of their specific course of study, took an identical curriculum during their freshman and sophomore years. Specialization occurred during the junior and senior years, with civil engineering students required to work their way through the following curriculum:

Junior Year

Fall Session: Rational Mechanics; Differential Calculus; Descriptive Geometry; Surveying; Graphical Statics.
Practicum: Surveying; Mechanical Drawings; Descriptive Geometry; Experimental Mechanics.

Winter Session: Applied Mechanics; Graphical Statics; Railroad Surveying and Earth-work; Integral Calculus; Spherical Projections and Map Drawings.
Practicum: Plotting Field Notes; Map Drawing; Physics.

Spring Session: Resistance of Materials; Shades and Shadows and Isometric Projection; Physics; Mineralogy; Civil Government.
Practicum: Experiments on Resistance of Materials; Mineralogy; Leveling and Topography.

Senior Year

Fall Session: Civil Engineering; Mental Philosophy; Geodesy and Practical Astronomy; Geology; Roads, Railroads, and Canals.
Practicum: Geodesy.

Winter Session: Specifications and Contracts; Geology; Political Economy; Hydraulics, Water Supply and Drainage; Civil Engineering.
Practicum: Drawing.

Spring Session: Rivers and Harbors; Astronomy; Civil Engineering and Engineering Projects; Review; Ethics.
Practicum: Field Practice; Thesis.

Old Main as it appeared from its completion in 1863 until renovated in the mid-1890s. (Penn State Collection)

John F. Healy was the first graduate of the civil engineering course, receiving a Bachelor of Science degree in 1884.

Except for a few of the practicums, all engineering classes—indeed, nearly all College classes—met in the Main Building. That structure, the original Old Main, had since 1859 served as Penn State's sole classroom building, as well as a dormitory for students, a residence for some of the faculty, and the hub of social activities.

Once the civil engineering program had taken root, President Atherton turned his attention to developing additional engineering courses that would provide greater diversity of instruction. In the spring of 1883 he approached young Louis E. Reber, an instructor in mathematics and temporary instructor in military tactics, with a startling proposition. Reber, born and raised in Centre County, was an 1880 graduate (and class valedictorian) of the College. Dissatisfied with the low pay and apparent stagnation of academic life, he was preparing to move to Texas, where he planned to enter private business. As an alternative, Atherton proposed that he spend the next year at the Massachusetts Institute of Technology, taking graduate work in mechanical engineering and paying particular attention to the methods of engineering education. Atherton promised to delegate to Reber upon his return to Penn State responsibility for restructuring the existing two-year program in mechanic arts and developing a four-year baccalaureate curriculum in mechanical engineering. Reber quickly agreed to the President's challenge, in spite of the fact that he would have to pay his own way at M.I.T. It was one of the most fateful decisions in the history of engineering education at Penn State.

Prior to returning to his alma mater in the fall of 1883, Reber took advantage of the summer recess to visit other engineering schools and observe first-hand their mechanical engineering curriculums. Institutions on his travel agenda included Massachusetts' Worcester Polytechnic Institute, Stevens Institute of Technology in New Jersey, Washington University in St. Louis, and the University of Minnesota. The mechanical engineering programs at these schools must have seemed grand in comparison to the limited offerings and lack of facilities found at The Pennsylvania State College, whose mechanic arts course consisted mainly of instruction in mechanical drawing, carpentry, and woodworking. Drawing was taught in a classroom in the Main Building, while the attic of the College pump house was converted to a carpentry shop. Reber supervised the installation of a makeshift forge and foundry in the pump house itself but could do little else without more spacious quarters. Late in 1884 he asked for $3500 for construction of a new building to be devoted exclusively to the work of mechanic arts. The request was not immediately approved, owing to a lack of state appropriations. Then in 1885 the College's Department of Agriculture reported a small surplus of funds arising from money Professor Whitman H. Jordan had earned by performing some fertilizer analyses for the Commonwealth. President Atherton directed that these funds ($1650) be combined with a small amount of available cash and used to finance construction of a mechanic arts building. The new structure—the first to be erected at the College for strictly academic purposes—was a modest 34' × 50' two-story, wood-frame affair located adjacent to the pump house (on the approximate site of the east end of today's Hammond Building).

To further stretch the limited dollars allotted to his department, Reber hit upon the idea of cajoling various equipment manufacturers to supply furnishings for the new building at reduced cost. Emphasizing the advertising value of such an arrangement, Reber managed to purchase for the College several instruments at prices well below commercial levels. These included a 16" turret lathe from Pratt and Whitney of Hartford, Connecticut; a 36" drill press and a large planer from William Sellers and Company of Philadelphia; and a shaper from Browne and Sharpe of Providence, Rhode Island. Into the Mechanic Arts Building went these and a host of other practical appliances, so that in a short time the structure contained a carpentry shop, a wood turning room, a forge room, and a machine shop. "The Pennsylvania State College now had shops, though small, as well equipped as any I had visited," Reber later recalled.[11] Total purchase price for all this equipment was only $2000.

The mechanic arts course was intended for those students who wished to acquire a high degree of manual training. Those who desired to learn the theoretical aspects of mechanics could enroll in the new mechanical engineering course, approved by the trustees for the 1886–87 academic year. Two rooms in the basement of the Main Building were outfitted as laboratories for the new Department of Mechanical Engineering, and a room on the first floor was used for lectures and recitations. As head of this department, Louis Reber was also assigned to oversee the operation of the pump house, whose steam-driven pumps forced water from a nearby artesian well into a small reservoir ensconced on the hill in the northwest sector of the campus; he also managed the steam heating plant in the basement of Old Main. When incandescent lighting was introduced into that building in 1887, the mechanical engineering department supervised the operation of the 50-horsepower steam engine and accompanying generator that consti-

Mechanic Arts Building, 1886. The pumphouse is at right, College Avenue in the foreground. (Penn State Collection)

tuted the electric light plant. Those chores were at times burdensome, yet the facilities did provide a fine opportunity for engineering students to gain practical experience.

By the time the course in mechanical engineering appeared, all students no longer took an identical course of instruction for the first two years. Instead, students were segregated according to their "general" or "technical" aspirations. A student in mechanical engineering in the late nineteenth century had to work his way through the following typical curriculum:

Freshman Year

Fall Session: Algebra; Geometry; German; History.
Practicum: Drawing, Geometrical and Projection; Carpentry.

Winter Session: Trigonometry; Geometry; Rhetoric; German.
Practicum: Drawing, Intersections; Carpentry.

Spring Session: Trigonometry; Physiology; German; Military Tactics.
Practicum: Drawing, Intersections and Developments; Wood-turning.

Sophomore Year

Fall Session: Analytical Geometry; Chemistry; German; French; History; Surveying.
Practicum: Surveying; Chemistry.

Winter Session: Analytical Geometry; Chemistry; German; French; History.
Practicum: Chemistry; Pattern-making.

Spring Session: Chemistry; French; Differential Calculus; Descriptive Geometry; Elements of Mechanisms.
Practicum: Chemistry; Drawing, Descriptive Geometry.

Junior Year

Fall Session: Physics, Mechanics, and Heat; Descriptive Geometry, Maps, Shades and Shadows; Integral Calculus; Mechanical Movements.
Practicum: Mechanics; Mechanical Drawing.

Winter Session: Analytical and Graphical Statics; Physics; Materials of Engineering; Differential Equations; Valve Gearing.
Practicum: Physics; Forging.

Spring Session: Kinetics and Kinematics; Thermodynamics; Materials of Engineering; Physics.
Practicum: Chipping and Filing; Mineralogy.

Senior Year

Fall Session: Determinants; Mechanics of Machinery; Geology; Political Economy.
Practicum: Mechanical Drawing: Engine Lathe Work.

Winter Session: Quarternions; Steam and Steam Engines; Indicators, Injectors, and Governors; Constitutional Law; Astronomy.
Practicum: Mechanical Drawing; Machine Construction; Testing Strength of Materials.

Spring Session: Machine Design; Quarternions; Hydraulic Motors; International Law.
Practicum: Machine Construction; Thesis Work.

The first graduates of the mechanical engineering course, all of whom received their degrees with the class of 1889, were John Price Jackson, H.E. Miles, and Jacob Struble.

Responsibility for instruction in the new but rapidly growing field of electrical engineering initially belonged to the Department of Physics, headed by I. Thornton Osmond. Upon his recommendation and the subsequent approval of President Atherton, the department greatly enlarged its work in the practical applications of electricity and in 1887 changed its name to the Department of Physics and Electrotechnics. The evolution of electrical engineering at Penn State thus mirrored a nationwide trend that saw empirical studies in electricity emerge first in physics departments and later become distinct academic disciplines.[12] Unlike the fields of mechanical and civil engineering, where earlier practical experience played a dominant role in shaping the curriculum, science preceded art in electrical engineering; that is, scientific discoveries in electricity and magnetism provided the knowledge that made possible applications-oriented studies. Electrotechnics held a special appeal for Penn State students in this, the dawn of the age of electricity. In spite of a lack of sufficient space and adequate laboratory facilities, within a few years the Department of Physics and Electrotechnics became one of the largest in the College.

In fact, all three engineering curriculums were proving extremely popular. Of the 92 degree students enrolled at the College in 1887–88,

Louis E. Reber (right rear, wearing derby) with mechanical engineering seniors, 1890. (Penn State Collection)

18 were studying mechanical engineering and 15 civil engineering. The following year total enrollment rose to 113, while the number of engineering students climbed to 48 (22 mechanical, 17 civil, and 9 physics and electrotechnics). President Atherton disclosed with considerable satisfaction in his annual report for 1887 that instruction in engineering "has been made more definite and systematic, and the several allied courses of mathematics, civil engineering, mechanical engineering, and physics thoroughly meet the requirements of the law of Congress, that one of the leading objects of the College should be to teach such branches of learning as are related to the mechanic arts."

To accommodate the growing undergraduate population, Atherton appealed to the General Assembly for $100,000 for a new engineering building. (The Morrill Act prohibited schools from using their endowments to erect buildings.) His request was initially denied, although the Commonwealth did appropriate a large sum of money for a chemistry and physics building and an armory. Atherton renewed his plea the next year. "The Departments of Mechanical and Civil Engineering, and the elementary Department of Mechanic Arts are so greatly overcrowded as to interfere with their work," he told the legislature. "Nothing but the zeal and energy of the professors in charge, combined with the earnestness of the students in these departments, has prevented the work from suffering greatly on this account."[13]

To assist in convincing state officials of the need for more money, Louis Reber prepared for President Atherton a card file containing the names of all senators and representatives, their hobbies, business interests, educational background, close friends, and political leanings. Reber then made sure that every legislator was presented with a copy of the College catalog by an alumnus who lived in his district, who also took the occasion to inform the lawmaker of the good work of the College and its need for increased financial support. Both Reber and Atherton understood that the vast majority of legislators were practical men and were most likely to be influenced by a glimpse of the kind of utilitarian education the College had to offer. Visits by alumni were therefore followed by invitations to inspect the campus, where the politicians toured the engineering facilities under the personal guidance of Professor Reber.

Atherton and Reber's cultivation of good relations with the legis-

Dinner in the Armory following dedication of the Main Engineering Building. (Penn State Collection)

lators paid dividends. Early in 1891 the Commonwealth allotted $100,000 to the College for a building to house civil and mechanical engineering departments. Ground was broken in June 1891 and less than two years later, on February 22, 1893, the new edifice, designed by College architect Fred L. Olds, was dedicated. Three stories high with a single-story rear wing and composed mostly of red brick, the Main Engineering Building, as it was christened, occupied a site not far from the main campus gate (about where Sackett Building now stands). The dedication was one of the grandest ceremonies staged at the College and deservedly so, since the new building was the most expensive yet erected, with the exception of Old Main. Nearly three hundred invited guests gathered in the middle of a howling snowstorm to hear President Atherton, Governor Robert E. Pattison, U.S. Secretary of the Interior John W. Noble, M.I.T. President Francis A. Walker, and a host of other dignitaries expound on the progress of engineering education at Penn State. After inspecting the building and lunching at the Armory, Atherton and his guests adjourned to Old Main's chapel for the obligatory speechmaking. General James A. Beaver, a trustee and chairman of the building committee, spoke first. Remarking that construction of the Main Engineering Building symbolized "a step out of the old into the new," Beaver pointed out that, "There is a university work which is outside the line of the old university and which is directly and distinctly in the path which this institution has marked out for itself. . . . The aim and outcome of the new university is the Engineer." Turning to President Atherton, Beaver officially transferred the building to the College, saying,

> We hand you this building over, in the expectation that the trustees will see to it that there is such equipment in the near future as will meet the demands of this age and will satisfy the wants of the community and will answer the question of the future, the question of today: "Where is the Electrician? Where is the Engineer?" We must have them both, and we depend in Pennsylvania, to a large extent, upon this college to furnish them.

Taking the podium briefly, Atherton thanked General Beaver, the other trustees, and the Commonwealth. Then, in a remarkable statement relevant for later times as well as his own, the president declared that the College and society at large should not look upon technical education as a panacea for the nation's ills.

> I hear a louder cry than that which has been named. I, too, hear the coming century asking for the Engineer, the Electrician, . . . but, when I

see a Republic like ours, whose foundations are and must be laid in the intelligence of the people, when I see in the midst of it this seething ferment of unrest that is taking possession of the hearts and minds of men, I conceive that one of the highest, if not the very highest, duties of an institution of this kind is to train men for citizenship, and to me the cry of the coming century is, "Give me the Engineer, Give me the Electricians," but above all, "Give me the all-around, well-trained man."

In his turn, Governor Pattison presented a history of higher education in Pennsylvania and its relationship to the state government and concluded by affirming, "We never were so dependent upon the educational interests in our land as we are today!"[14] Still more speeches followed, until at length the final syllable was uttered, and the crowd dispersed for an evening of socializing.

Not present on the speakers' platform that cold winter day was the man who as much as any single individual was responsible for the fine laboratory equipment with which the building had been furnished. To Louis E. Reber, although head of only the Department of Mechanical Engineering, Atherton had delegated the job of securing the hardware necessary for the students' practicum work. Reber had for some time desired to purchase a Reynolds-Corliss triple-expansion steam engine for the College. (He had been captivated by this type of machine ever since viewing the famous Corliss engine on display at the Centennial Exhibition in Philadelphia in 1876.) Since the new engineering building would provide an ideal setting for such a machine, he took the matter up with President Atherton. Atherton, his enthusiasm for engineering education notwithstanding, balked when Reber explained that the engine would cost about $10,000. Reber persisted. "If we purchase this machine it will equal the best in the country," he argued. "Can The Pennsylvania State College afford to be inferior in this respect?" The appeal to Atherton's pride succeeded, and Reber was authorized to acquire an engine as soon as possible. He journeyed to Milwaukee, Wisconsin, home of the Allis-Chalmers Company, the engine's manufacturer. The firm responded enthusiastically, even suggesting several changes in the engine which, while adding to its expense, would better it for the experimental work of students. Allis-Chalmers then agreed to sell the engine—modified according to Reber's specifications—to the College for $8000. This sum was considerably below the $10,000 which an unmodified engine would have brought on the commercial market.[15]

Assisting Professor Reber in procuring the needed laboratory materials was Professor Barnard and his successor (in 1893) as head

Main Engineering Building, as viewed from the bell tower of Old Main. (Penn State Collection)

of the Department of Civil Engineering, Fred E. Foss. Machinery and other gear came from sources as varied as the Westinghouse Air Brake Company, the Detroit Lubricator Company, the American Steel Packing Company, and the Pennsylvania Railroad. Reber recollected later that, "It was not difficult to complete the laboratory equipment, though some of the machines were expensive. Those purchased were the best of their kind and were secured at a cost considerably below the commercial rate in view of student use."[16]

The Main Engineering Building had originally been intended for the use of the civil and mechanical engineering departments. However, the trustees approved a course in mining engineering to begin in the fall of 1893, and the new Department of Mining Engineering was to have its quarters in the new building, too. Magnus C. Ihlseng, formerly of the Colorado School of Mines, was named Professor of Mining Engineering and head of the department. At the same time the trustees agreed to the establishment of a mining engineering curriculum, they also consented to the separation of the Department of Physics and Electrotechnics into two distinct entities. Professor Osmond continued to head the Department of Physics, while John Price Jackson was appointed head of the new Department of Electrical Engineering, one of fewer than a half-dozen such departments in all of American higher education. John Price Jackson was the son of Josiah Jackson, the College's Professor of Mathematics. The younger Jackson had been among the first students to enroll in Reber's mechanical engineering course and had been one of the first graduates. In 1888 he and his former mentor became brothers-in-law when Reber married Jackson's sister, Helen. After graduation he had worked for a time in the electric railway field for the Sprague Electric Company and its successor, General Electric. In 1891 he returned to his alma mater to become Assistant Professor of Electrical Engineering in the old physics and electrotechnics department. Now, at age 24, Jackson was easily the youngest department head at the College.

The administration had little alternative but to house the Department of Electrical Engineering in the Main Engineering Building. Within a few months after that structure's dedication, four departments were squeezed into space initially designed for only two. Compounding the problem was the unceasing rise in enrollment in all four departments. The overcrowded conditions caused President Atherton in 1894 to request money for the erection of an additional building for the civil and mining engineering departments. Even if only three, or two, of the engineering departments were located in the Main Engineering Building, sufficient space could not have been guaranteed, given the burgeoning number of students who wished to become engineers. In 1890–91, 127 undergraduates were in attendance at the College. Seventy-three were enrolled in engineering courses, of whom 37 were in civil, 19 in mechanical, and 17 in electrotechnical. In 1893–94, the total student population climbed to 181, with 128 enrolled in engineering (57 in electrical, 44 in mechanical, 18 in civil, and 9 in mining). Engineering enrollment showed a 75 percent increase in this three-year period, as opposed to a 35 percent increase in total enrollment. Engineering students comprised over 70 percent of the student body by 1893.

The engineering profession had entered its golden age. Not only were engineers visible contributors to the nation's economic growth; the technological progress that they had brought about promised to lead to a material equality that was in every way a fitting complement to the political and social equality that Americans liked to believe were hallmarks of their society. Engineers were heroes in an age in which the public, not yet satiated by technological marvels, could still wax enthusiastic over the completion of a giant suspension bridge or the installation of a network of electric street lights. And were not engineers improving society by their way of thinking as well as by their deeds? Could not other persons benefit by applying the same kind of objective, open-minded scientific methods that engineers used? Little wonder that so many youths wished to become engineers, the high priests of the new epoch.

The curriculum emphasized general training in all the major elements of a particular engineering discipline, as opposed to specialized studies concentrating on only one or two of these elements. Students in mechanical engineering, for example, studied steam engineering (including reciprocating engines, boilers, and calorimetry), hydraulic motors, heating and ventilation, machine design, and friction and lubrication. As late as 1894 Louis Reber taught most subjects, with the exceptions of machine design and mechanical drawing. Lieutenants John Pemberton, and later Thomas W. Kinkaid, both on loan from the Department of the Navy, taught these. (The Navy underwent a sizeable reduction in equipment and personnel in the latter half of the nineteenth century. Rather than discharge many of its highly trained officers, the government detailed them to the land-grant colleges, where their technical expertise could be put to good use.) Supervising the mechanics shops for many years was Assistant Professor William M. Towle, an alumnus of the Worcester Polytechnic Institute, the school that originated shop work for engineering students. Under Towle, students learned many of the practical elements of mechanical engineering and acquired a degree of manual dexterity in carpentry, foundry and forge work, and tool and pattern-making. Students in the two-year mechanic arts course spent nearly all their study hours in the shops, for their curriculum gave priority to the inculcation of manual skills. In many ways the mechanic arts course represented a lingering remnant of a bitter controversy that raged among engineering educators during the 1870s and 1880s, the point of contention being how much of the college curriculum should be devoted to abstract principles and how much to specific applications. Penn State had not participated in most of the debate. By the time its engineering programs emerged, the theorists had triumphed over the vocationalists, and Penn State engineering evolved along lines similar to those of most other engineering schools. A great deal of sentiment remained for vocational training, nevertheless; hence, the retention of the mechanic arts course.[17]

Electrical engineering was by its very nature more scientifically oriented, yet even students in that curriculum still received a great deal of training directed toward practical applications, as this typical curriculum from the 1890s demonstrates:

Freshman Year

Fall Session: Algebra; Solid Geometry; English; French or German.
Practicum: Freehand and Geometric Drawing; Carpentry.

Winter Session: Trigonometry; English; French or German; Military Tactics.
Practicum: Projection Drawing; Carpentry.

Spring Session: Analytic Geometry; English; French or German; Ancient History.
Practicum: Machine Design; Wood Turning.

Sophomore Year

Fall Session: Analytic Geometry; French or German; Chemistry; Medieval History.
Practicum: Machine Design; Chemistry; Forging.

Winter Session: Engineering Mechanics; Calculus; French or German; Chemistry; Modern History.
Practicum: Machine Design; Chemistry; Chipping and Filing.

Spring Session: Engineering Mechanics; Calculus; Descriptive Geometry; Physics.
Practicum: Descriptive Geometry Drawing; Chemistry.

Junior Year

Fall Session: Engineering Mechanics; Calculus; Differential Equations; Physics.
Practicum: Physical Measurement; Moulding.

Winter Session: Engineering Mechanics; Hydraulics; Steam Engines; Dynamo Machinery; Physics.
Practicum: Electrical Measurement; Lathe Work.

Spring Session: Boilers, Engineering Materials; Lubricants; Metallurgy; Primary Batteries; Dynamo Machinery.
Practicum: Electrical Measurements; Calibration of Instruments; Testing Materials.

Senior Year

Fall Session:	Thermodynamics; Electric Lighting; Telephone and Telegraph; Economics. Practicum: Engine and Boiler Tests; Dynamo Tests.
Winter Session:	Hydraulic Motors; Railways and Transmission of Energy; Installation and Machine Design; Constitutional Law. Practicum: Electrical Engineering Laboratory; Installation and Machine Design.
Spring Session:	Electricity in Mines; Electro-Metallurgy; Specifications and Estimates; International Law. Practicum: Thesis.

Arthur G. Guyer and John B. Hench were the first students to be awarded degrees in electrical engineering. Both received Bachelor of Science degrees at the 1894 commencement. To ease the load on Professor Jackson, who heretofore was the sole member of his own department, Henry A. Lardner was appointed Instructor in Electrical Engineering in the fall of 1894.

Under Professor Barnard, the Department of Civil Engineering made great strides during its first decade of existence. Instruction in sanitary engineering and hydraulic engineering was expanded to give students stronger preparation in these fields. Perhaps the most popular subjects in the department at this time related to railway surveying and construction, reflecting the railroad mania that was then sweeping the nation. As in mechanical and electrical engineering, civil engineering students did not yet have the opportunity to specialize in one particular facet of their profession. Their only chance to pursue topics of special interest to them came during laboratory work and when researching and writing their senior theses. Just when the Department of Civil Engineering moved into its new building and seemed to be on the verge of further expansion, Louis Barnard resigned his post and moved to England to join relatives there. Named as the new head of the department was Fred E. Foss, a graduate of Bates College (Maine) and the Massachusetts Institute of Technology. T. Raymond Beyer served as Instructor and later Assistant Professor of Civil Engineering under both Barnard and Foss.

Beginning in 1894, all freshman, sophomore, and junior engineering students were required to take a two-week summer course immediately following the close of the spring term. During these sessions (the first summer sessions in Penn State's history), students gained field experience that for one reason or another the departments were unable to furnish during the regular academic year. The summer course also featured visits to coal mines, railroad shops, foundries, power stations, and other installations where students could observe firsthand the things they had previously known only from textbooks and lectures.

In his annual report for 1890, Louis Reber proudly noted that all eight graduates of the mechanical engineering course that year had secured jobs well before the June commencement, and all had won at least one promotion before the year was out. These bright employment prospects extended to graduates of all the engineering courses, as the demand for persons having extensive technical training outstripped the supply. Young engineers experienced little difficulty in achieving positions of considerable responsibility, and many went on to attain national and even international prominence. Cummings C. Chesney ('85, physics and electrotechnics), for example, became a partner of the famed electrical genius William Stanley and helped develop an efficient system for the transmission of polyphase alternating current that made possible the introduction of low-cost electricity into millions of

Civil engineering students measuring the velocity of a stream near Thompson's Spring, east of the College. (Penn State Collection)

homes across the country. Another physics and electrotechnics graduate, Frederick Darlington ('85), worked for many years as a nationally renowned consultant in the electric railway field before directing a federal government study of the feasibility of interconnecting utilities, a study that paved the way for today's giant power pools and grid networks. Jacob Struble ('89, mechanical engineering) and James C. Mock ('90, mechanical engineering) were leaders in the perfection of automatic block signals, a major advance in railway traffic control and safety. Struble spent most of his years with the Union Switch and Signal Company, while Mock left that firm after a few years to supervise pioneering signal installations for the Pennsylvania, New York Central, and other railroads. Engineers who aspired to managerial positions in business and industry found their technical training to be a valuable asset. Philip G. Gossler ('90, mechanical engineering) rose from draftsman to president of the Columbia Gas Company. C.C. Chesney retired as a vice-president of the General Electric Corporation, and Frederick Darlington ended his career as a vice-president with the Westinghouse Electric and Manufacturing Corporation. Arthur G. McKee ('91, mechanical engineering) headed his own consulting and contracting firm, the Cleveland-based Arthur G. McKee and Company, that did hundreds of millions of dollars worth of business worldwide.

A few students who spent four years in pursuit of an engineering degree later chose to enter a different profession. Perhaps the most notable in this category was J. Franklin Shields ('92, civil engineering), who taught mathematics briefly before entering law school. He became a distinguished lawyer in his native Philadelphia and ultimately served a lengthy term (1929–46) as president of Penn State's board of trustees. Certainly the most unusual case of a non-practicing engineer involved Carrie McElwain ('93, civil engineering), younger sister of the College's Lady Principal, Harriet McElwain, and the first female to earn a Penn State engineering degree. Female undergraduates had attended the College since 1871, but most had enrolled in a special "ladies' course" or had taken one of the general courses. Why Carrie McElwain chose to study engineering in an era when nearly any kind of technical training was considered unsuitable for women remains something of a mystery. She married classmate Edward P. Butts ('93, civil engineering) shortly after graduation and is not known to have ever worked in an official capacity as an engineer.

Not to be overlooked are engineering alumni who remained in academia. Foremost among these was Dugald C. Jackson ('85, civil engineering), older brother of John Price Jackson. After completing graduate studies at Penn State and Cornell University, Jackson worked for several large electrical equipment manufacturers before moving to yet another land-grant institution, the University of Wisconsin, where he organized a Department of Electrical Engineering in 1901. Several years later he accepted an appointment as head of the electrical engineering department at the Massachusetts Institute of Technology. In addition to heading a consulting firm of international repute, Jackson was the first Penn State alumnus to be elected president of the Society for the Promotion of Engineering Education (original name of today's American Society for Engineering Education) and the American Institute of Electrical Engineers.

Formation of the School of Engineering

The College was able to finance the expansion of engineering education in part because of a supplement Congress attached to the original Land Grant Act. The Second Morrill Act, passed in 1890, appropriated an initial $15,000 to each land-grant institution. A thousand dollar increase in this sum followed annually until a maximum of $25,000 was reached. The money was to be spent on instruction in "agriculture, the mechanic arts, the English language, and the various branches of the mathematical, physical, material, and economic sciences, with specific reference to their applications in the industries of life." The extra funds thus acquired were especially welcome in the absence of tuition fees. Penn State had abolished all tuition (except for music instruction) in 1874 and levied only nominal charges for heat and light expenses and for the use of the laboratories. Yet the school entered the decade of the 1890s on a more sound financial footing than it had enjoyed in any previous period, thanks to the land-grant endowments and the increased size of appropriations from the state legislature.

President Atherton professed to see a negative aspect to this good fortune, however. He maintained that the expansion of the technical courses, primarily those in engineering, absorbed so much of the College's income that they inhibited the growth of the non-technical courses. In his annual report for 1894, for example, he stated:

> The technical departments have grown so rapidly, and have so fully absorbed all our available resources, that it has been impossible to maintain the general courses with the same breadth and efficiency. Not all students wish to become engineers. . . . If we cannot enlarge our work in *all* directions, I am clearly of the opinion that we should hold down our technical work at the present level.

Resources conserved by this retrenchment Atherton hoped to channel toward broadening and strengthening the liberal arts. In the early years of his administration, the president had wisely assigned priority to nurturing engineering studies in order to bring the College into compliance with the Morrill Act and to win the confidence of the citizens of the Commonwealth and their leaders, who were most likely to be impressed by the utilitarian functions of higher education. By the mid-1890s, Atherton appeared to have succeeded in reaching these objectives. Now it was time to bolster the liberal arts. The president was convinced the College should upgrade its general education studies, not only as a service to those students who had no desire to pursue technical careers, but also for students enrolled in the various engineering and science courses. Most of the engineering faculty shared Atherton's belief that the technical curriculums must embody a significant number of liberal subjects. Typical was the view of Louis Barnard, who remarked that the engineering departments' greatest challenge did not lie in technical training itself, but rather in "how to adjust the demands of the technical training in such a manner that the student may not only become a successful engineer, but a man of broad culture and a good citizen as well."[18] Instruction in one or more of the liberal arts had been offered since the College's founding in 1859, yet except for English, no one subject had been taught continuously since that time. Throughout 1895, President Atherton worked with the faculty to determine ways in which past gains in technical education could be consolidated and the path cleared for improved instruction in non-technical areas. The fruits of their labors became apparent late in the year, with the grouping of all departments into seven schools: Agriculture; Engineering; History, Political Science, and Philosophy; Language and Literature; Mathematics and Physics; Mines; and Natural Science.

The creation of schools placed the technical and the general courses on a more even plane, although Atherton was the first to admit that technical instruction still ranked higher, and deservedly so, considering the College's obligations as a land-grant institution. The establishment of schools had another, equally important objective: to encourage and facilitate cooperation among related departments. In this way duplication of instruction and departmental resources could be avoided, while lines of communication between departments would be made more effective. Furthermore, much of the burden of administration was shifted away from the president's office and onto the new deans, as the deans would henceforth be responsible for such items as scheduling classes and instructors and administering discipline to any

Louis E. Reber (Penn State Collection)

student who failed to meet academic requirements. To the engineering faculty, at least, the organization of a School of Engineering was a most welcome measure. Hardly an annual departmental report had been written over the last five years that did not contain a plea for increased interaction among the separate academic units. Neither engineering subject matter nor student interests recognized the artificial boundaries delineating the departments. Placing the departments under a central administration would eliminate wasteful overlapping and provide better guidance in planning for future expansion of engineering education, which at this point was deemed to be a virtual certainty.

To be Dean of the School of Engineering, Atherton selected Louis Reber. The choice was a tribute to the years of dedicated service and strong leadership Reber had compiled at Penn State. He had several years before turned down an offer from the Ohio State University to become head of that institution's mechanical engineering department, even though the position carried a salary considerably higher than the $2000 that he was earning at Penn State. Students greeted his appointment with mixed feelings, however. "Ike" Reber, as they affectionately called him, was an extremely popular and effective teacher, and they knew that his new administrative chores, together with his duties as head of the Department of Mechanical Engineering, would greatly reduce the amount of time he spent in the classroom. The new Dean of Engineering exercised authority over the Departments of Civil, Mechanical, and Electrical Engineering. The Department of Mining Engineering became the core around which the new School of Mines was to be built. Magnus Ihlseng was named Dean of that school.

The board of trustees formally consented to the organization of the schools in January 1896, and the first phase of the history of engineering education at Penn State came to a close. The struggle to establish formal instruction in engineering had been a long and arduous one. Thanks to a group of concerned faculty and to a president who understood the importance of engineering, that struggle had ended in victory. Indeed, the popularity of the engineering courses probably more than any other single factor was responsible for the generous financial support the Commonwealth accorded the College in the late 1880s and early 1890s. That support, the land-grant income, and the high caliber of leadership displayed by faculty and administration was quietly propelling The Pennsylvania State College upward in the ranks of the nation's engineering schools. Dean Reber stated with pride in his annual report for 1895 that, "The more closely I become acquainted with the equipment and work of some other large engineering schools, the better I am satisfied with the equipment provided and work being done at our own institution."

2 Growth and Maturation: 1895–1915

In 1895–96, the School of Engineering of The Pennsylvania State College enrolled 143 undergraduates, about two-thirds of the College's total student population. It was housed in the College's largest and most expensive building devoted exclusively to academic activities; its graduates were eagerly sought by some of the nation's wealthiest and most prestigious industrial firms; and its importance to the economic well-being of the Commonwealth was obvious. Yet the School had by no means solved all of its problems. The tribulations that it would encounter over the next decade or so were not so severe as those experienced in laying the foundation of engineering education; but they were formidable nonetheless and for a time seriously jeopardized the quality of Penn State's engineering programs.

Coping with the Demand for Technical Education

Ironically, the School of Engineering's troubles stemmed in large part from the very successes it had achieved. The size of Penn State's student body increased only marginally during the last few years of the nineteenth century. Beginning with the 1899–1900 academic year, however, the number of students attending the College and consequently those taking engineering courses climbed dramatically. By 1901–02, enrollment had more than doubled over the preceding three years, reflecting the expanding role college-trained engineers played in American life. On the other hand, state appropriations were not keeping pace with this increase. Between 1887 and 1897 the Commonwealth had allotted a total of $680,000 to the College. Approximately

$600,000 of this amount went for building construction and repair, leaving an average of $8000 per year to supplement the $65,000 or so the College annually derived from its land-grant endowment to purchase new equipment and pay salaries. This sum was fast becoming insufficient to meet the growing needs of the institution. And while the School of Engineering received a large portion of Penn State's total income, the amount was still pitifully inadequate and caused a variety of hardships for engineering faculty and students alike. Among the more serious difficulties plaguing the School was its inability to enlarge its teaching force to meet the larger student load. At its creation in 1895, the School of Engineering numbered nine faculty members (not including assistants) and, as mentioned, 143 students. In 1900–01, the School enrolled 237 students, but it still had a teaching staff of only nine. The appropriations of the General Assembly after 1897 simply did not permit the hiring of new faculty, except on a replacement basis. Nor was the School able to enlarge its laboratory facilities during this time, with two notable exceptions. The first of these was the Department of Electrical Engineering's experimental electric railway, which was installed under the supervision of John Price Jackson in 1896 and utilized the right of way of the Bellefonte Central Railroad. This 18-mile shortline linked the College with the county seat and the Pennsylvania Railroad at Bellefonte. The electrified segment began near the Bellefonte Central station behind the Main Engineering Building and terminated about a mile distant at the wye at Strubles. Several industrial firms donated most of the materials needed to construct the line. The department obtained a single-truck trolley car (probably a donation from the Philadelphia Traction Company) and propelled it with 500-volt direct current received from an overhead wire connected to a generator in the College power plant. Completed at modest cost, the experimental railway provided an excellent means for students to acquire experience in the field of electric traction, which in the form of street and interurban railroads was approaching the zenith of its popularity.

The second noteworthy addition to the School's laboratory resources belonged to the Department of Civil Engineering. In 1899 that department installed a fully equipped hydraulic laboratory in the main building. Having as its centerpiece an 82-foot standpipe that stretched from basement to attic, the facility equalled in nearly every respect the hydraulic laboratories found at the more richly endowed engineering schools.

The inability of Penn State's engineering departments to secure more equipment for practical instruction came at a particularly bad

ENGINEERING EDUCATION AT PENN STATE

Turn-of-the-century electrical engineering students finding the impedance of the secondary transformer. (Penn State Collection)

time, since the turn of the century witnessed many important technical advances in engineering materials and methods. In the field of electrical engineering, for instance, high-voltage alternating current was rapidly gaining ascendancy over the less efficient low-voltage direct current that previously enjoyed widespread use. Telephones were becoming household items, and phone lines crisscrossed even some of the most isolated segments of the countryside. Electric motive power appeared to be on the verge of supplanting coal-burning locomotives on some of the nation's largest steam railroads. With considerable justification, therefore, did John Price Jackson in 1901 demand that his department be allocated funds to outfit a high-voltage a.c. laboratory and a telephone laboratory, and to modernize the electric railway. The estimated costs of these "minimum" improvements was $34,000.

That same year Louis Reber requested a similar allocation for his Department of Mechanical Engineering. He desired to add a refrigeration test plant, a materials testing laboratory, and a full-size steam locomotive. He had already entered into negotiations with the Baldwin Locomotive Works of Philadelphia hoping to induce that firm to sell a suitable engine to the College at a discount price. Reber judged that these and other vital pieces of laboratory equipment would cost about $30,000.

Even if the needed apparatus could be procured, scant vacant space existed to accommodate it. The Main Engineering Building already bulged from cellar to rooftop with students and laboratory instruments. Dean Reber remarked with considerable understatement in 1901 that "the engineering building has become inadequate to the accommodation of the four departments at present housed within it."[1] He recommended that a two-story wing be attached to the rear of the building, extending in the direction of the railroad station. Mechanical and electrical laboratories and some of the mechanic art shops would occupy the new wing, thus freeing space in the main structure for classroom and office use.

President Atherton expressed sympathy for engineering's plight. He continually brought the crowded conditions and inadequate size of the faculty to the attention of the governor and legislature, but he met with little success in persuading them to grant the College a more generous appropriation. Furthermore, having awarded first priority for so many years to developing a sound curriculum in engineering, Atherton came under increasing pressure from the other schools to channel more of the College's income toward improving instruction in the sciences and agriculture. Atherton himself wanted to upgrade the liberal arts, his earlier plans to do so having been foiled by the College's budgetary difficulties. Reber recognized the detrimental effect this would have on the engineering courses and worked hard to retain what he considered to be his school's fair share of the money. Since engineering regularly enrolled about 70 percent of the student body, he pointed out to President Atherton:

> It does not seem unreasonable to ask that a proportionate amount of the available funds be devoted to this school. . . . I trust you will acquit me of a desire to belittle the work being done by other than engineering departments. I believe that I recognize the essential excellence of the work they are doing for this institution and appreciate the successes which are being achieved by their graduates; but I beg of you to bear in mind the fact that it is in the Engineering School that the larger numbers of our men are to be cared for and it is from the Engineering School that the larger numbers go out to make or mar our reputation in the world.[2]

Reber had good reason to fear the consequences of a reduction in the proportion of funds his school received. In 1900 the School of Mines reverted to the status of a department administered by the School of Engineering. Inadequate financial support and lack of space had seriously hindered the progress of the School of Mines since its formation.

In no area of technological activity was prejudice against college-trained engineers more deeply entrenched than in coal mining. Consequently, leaders in the coal industry made no concerted effort to persuade the Commonwealth to help sustain higher education in the mineral industries. Political considerations were often far more influential. Thus in 1895 the General Assembly allocated $50,000 for the creation of a mining school at the Western University of Pennsylvania, while appropriating a tiny fraction of that sum to assist a program in mining engineering already in existence at Penn State.[3] By the turn of the century, the College could no longer afford to maintain a separate School of Mines, particularly when enrollment showed no prospect of increasing. The mining school's extinction only intensified the already heavy drain on the School of Engineering's resources.

In addition to fulfilling his academic responsibilities, Louis Reber also supervised operations at the College's power plant, which provided steam heat to all the larger buildings on the campus and electricity (for lighting only) to all College buildings, the railroad station, the Beta Theta Pi fraternity house, and several professors' residences. Professor Jackson assisted Reber in overseeing operations and often used his electrical engineering students to extend and improve service throughout the campus area. Jackson was especially proud of the half-dozen arc lamps his students had installed to illuminate campus lanes and walkways. The arcs were on independent circuits, he reported, which meant that "they can be thrown on from one to two hours later than other circuits and need not be turned on at all on bright moonlight nights."[4] Such a pioneer energy-saving measure resulted in significant economies and was particularly welcome at a time of scarce dollars.

In 1900, in spite of the privations it had borne, Reber reported that Penn State ranked tenth in the nation in the number of undergraduate engineering students enrolled. The top ten schools and their enrollments were: Cornell University, 754; University of California, 729; Massachusetts Institute of Technology, 535; Purdue University, 450; Ohio State University, 330; University of Wisconsin, 328; Iowa State College, 311; University of Illinois, 287; University of Missouri, 214; and Pennsylvania State College, 209. Nine of these institutions were land-grant schools, and the tenth, the Massachusetts Institute of Technology, was a land-grant school in name if not in practice. Within the Commonwealth, Penn State enjoyed the largest engineering enrollment, with the University of Pennsylvania and its 161 undergraduates trailing a distant second. Nevertheless, Louis Reber could foresee the day when, owing to an absence of adequate finances, the College would drop from the ranks of the nation's largest technical institutions.

The College could not successfully meet the new challenges of engineering education by merely resting on past achievements. "If we do not steadily advance," Reber warned President Atherton in 1900, "there must be retrogression, a condition I do not believe that this institution can afford to permit in her Engineering School."[5]

Reber formulated a four-point program designed to avoid any retrograde movement. If implemented, his program would enhance the quality of instruction and diversify the curriculum to keep it abreast of the latest advances made in the various engineering fields. The dean proposed the establishment of four objectives for the School: a substantial increase in the size of the faculty; the acquisition of more and better laboratory equipment; a sweeping revision of the course offerings to permit greater specialization and to bring the curriculum into line with recent technological advances and the changing needs of large industrial employers; and the construction of more buildings to facilitate the attainment of the first three objectives.

Building construction entailed by far the most expense. Yet Reber was not so naive as to believe the legislature, which had in recent years shown no inclination to support even a modest expansion of engineering, could be expected to embrace the sizeable undertaking he now envisioned. Instead, he and Atherton believed they had found a new benefactor. Charles M. Schwab, multimillionaire steel magnate and a trustee of the College, in 1902 donated $155,000 for a new chapel, subsequently christened Schwab Auditorium. Reber and Atherton concluded that perhaps Schwab would be willing to underwrite the cost of enlarging the Main Engineering Building as well. After carefully inspecting the facility, Schwab remarked to Reber that "you certainly need larger and more commodious quarters," and asked the dean to send him detailed drawings of the new buildings he wanted.[6] With Atherton's blessing, Reber retained a Pittsburgh architect who in consultation with the engineering faculty prepared plans for two spacious extensions to the main building.

One of the three-story wings was to run adjacent to College Avenue. Measuring over 500 feet in length and 50 feet in width, it would house the Department of Electrical Engineering, new electrometallurgical and electric transportation laboratories, and a new campus power plant. (Coincidentally, the Hammond Building, constructed in 1960 on the site of the proposed new wing, measures approximately 600 feet in length and 60 feet in width.) The other wing, of about the same size, was to extend from the opposite or upper end of the main building and run parallel to the lower wing, thus forming a huge U-shaped complex. The second wing would be home for the Department of Mechanical

Engineering with its new steam, lubricant, gasoline engine, and railway mechanical engineering laboratories. The Department of Civil Engineering and the mechanic arts shops were to occupy the space in the old building vacated by electrical and mechanical engineering. Since mining engineering was expected to be reorganized again into a separate school, it would have to find quarters elsewhere. The two extensions, designed to harmonize with the romanesque architecture of the main building, carried a price tag of $920,000—$500,000 for the structures themselves and $420,000 for the equipment. Although this figure made the project the costliest physical expansion the College had yet contemplated, according to Reber it was not beyond the amount it was understood Schwab was willing to give. Before the architect finished his drawings, however, the steel titan suffered a series of financial reverses that he claimed were so acute that he had to postpone all philanthropic ventures. Shortly thereafter he left for an extended vacation in Europe. President Atherton, completed blueprints in hand, doggedly followed him across the Atlantic to beseech him to reconsider, but Schwab politely declined. In 1904, when Atherton finally recognized that no further aid would be forthcoming from Schwab, he appealed in vain to the General Assembly to appropriate at least enough money for the construction of a new building for the Department of Electrical Engineering. Meanwhile, as an emergency measure, the Department of Mechanical Engineering utilized cast-off materials and student labor to erect a small frame structure between the main building and College Avenue to be used as a foundry. The room formerly used as a foundry was converted to an alternating current laboratory.

Reber's efforts to increase the size of the teaching staff met with no better luck than his attempts to secure more buildings or laboratory equipment. In 1903–04, the engineering faculty (excluding assistants) numbered just 11 men: 5 in mechanical, 4 in civil, and 2 in electrical. Undergraduate engineering students in residence totaled 453. The departments did employ a number of teaching assistants, but these individuals were barely beyond their undergraduate days. Lacking practical experience and specialized knowledge, they could be of material aid only in introductory subjects and in shop work. Moreover, the faculty suffered from an extremely high turnover, caused primarily by poor salaries and by Penn State's geographic isolation. Teachers frequently remained on the staff only long enough to gain sufficient experience to enable them to obtain better paying positions at schools in more densely populated areas. The lure of private industry, which offered annual salaries two or three times greater than the $1500 or so

President Atherton (front row center, with white beard) and his department heads, 1903. To his immediate left is Louis Reber. Other Engineering department heads are Fred Foss and John Price Jackson (extreme left and right, respectively, in second row) and Marshman E. Wadsworth (behind Reber, with white beard). (Penn State Collection)

that engineering professors received, also contributed to the personnel problem. In return for their meager compensation, these men worked under conditions that bordered on the intolerable. Often one professor was forced to teach two separate classes simultaneously, sometimes in separate rooms, sometimes jammed into a single classroom. Laboratory equipment, worn from overuse and deferred maintenance, had the annoying habit of malfunctioning at precisely the time it was most needed. Classes were squeezed into every available square inch of the main building, including the halls. Reber surveyed the number of engineering students per instructor at The Pennsylvania State College in 1906 and found it to be twice as great as at the Massachusetts Institute of Technology, 2.5 times greater than at the University of Michigan, and 4 times greater than at Columbia University.[7]

Ironically, Dean Reber did win a modest degree of success in his attempt to provide more specialized instruction. In 1902–03 engineering electives appeared in the College catalog for the first time when the Department of Electrical Engineering offered its students the opportunity to take one of four options: general electrical engineering, electric railways, electrical energy (applications of electricity to industry), and electrochemical engineering. All students received the same basic instruction during their freshman and sophomore years. During their junior and senior years they took more specialized classes according to the option they had selected. In 1902, in conjunction with the creation of these options, the department acquired its first new pieces of laboratory equipment in six years: a rotary converter, an induction motor, and some measuring instruments. The general option was dropped the following year in favor of one in "Light, Power, and Signal Engineering." This option soon attracted a variety of students who were preparing to enter the commercial power, telephone, and railway signal industries. It proved less than satisfactory because of this diversity, and in 1906 the department once again restructured its options. Electrochemistry was elevated from an option to a full-fledged baccalaureate course with two options of its own, electrical and electrometallurgical. Four options remained in the electrical engineering course. These were: general electrical engineering, electric railway engineering, electrical applications (in industry), and telephone engineering. The electric railway option, although a popular one, apparently experienced more than its share of difficulties. In his annual report for 1903–04, John Price Jackson announced that the motor, truck, and controller of the trolley car had to be put into storage for lack of a suitable car body in which to house them. (Why the original car was no longer serviceable is not known.) In addition, the length of the electrified line was cut

Mechanical engineering students making boiler tests in the College power plant. (Penn State Collection)

back to just a half-mile. A new car was obtained in 1905 from the United Railways and Electric Company of Baltimore, but it does not seem to have been in use for more than a few years.

Reber's Department of Mechanical Engineering followed its counterpart in permitting greater specialization. The 1903–04 College catalog listed four options of study formulated "to meet the special requirements of those students who have decided to prepare for specific lines of work." These options were: general mechanical engineering, railway mechanical engineering, chemical and metallurgical engineering ("to prepare students for those lines of engineering which are concerned with the iron and steel industries"), and electromechanical engineering ("for students desiring to take up the special study of power plants"). Studies already introduced in electrical engineering nullified the need for an electromechanical option, so it was dropped from the curriculum in 1906. As in the case of electrical engineering, the freshman and sophomore years in mechanical engineering made no distinction as to option. Electives were available only to juniors and seniors.

Since becoming dean, Louis Reber had been an enthusiastic proponent of developing a strong course in railway mechanical engineering. He sensed that the railroad industry was entering a new era of steam motive power development, with larger and more powerful locomotives superseding the smaller designs that had been the rule for half a century. Railroads enjoyed unprecedented prosperity and needed thousands of expertly trained engineers, not only in the motive power field but also in nearly every other mechanical area as well. Reber very much wanted his students to lead this technological advance within the railway industry. In fact, at Altoona, a scant forty miles from the College, was situated the world's largest railroad shop complex, owned by the largest potential employer of Penn State railway mechanical engineering graduates. "The Pennsylvania Railroad Company is drawing a large number of its men from an institution whose reputation for engineering work is not so high as that of The Pennsylvania State College, but which is especially equipped for railway mechanical engineering work," complained Reber to Atherton in 1902. "Here is an unfortunate condition. Our own institution should be so well equipped in this line as to hold its own against any institution in the country."[8]

A year later, after the trustees had approved the division of the mechanical engineering course into the four options noted above, Reber's department secured permission from the Bellefonte Central to offer a practicum in "locomotive running." The practicum, usually taken by seniors, was intended to enable students to gather research material for use in preparing their theses. Various experiments were carried out over the entire length of the line between State College and Bellefonte utilizing Bellefonte Central locomotive No. 4, a small 2-8-0 type. Reber regarded the arrangement strictly as an interim measure. With Arthur J. Wood, newly appointed (1904) Assistant Professor of Experimental Engineering, he began negotiating with the Pennsylvania Railroad in hope of obtaining a locomotive and other gear the College could really call its own.

The Department of Civil Engineering joined the other departments in modifying its course of study to encourage specialization. In 1904–05, two options appeared in the catalog: general civil engineering and sanitary engineering. Sanitary engineering had taken on considerable importance as more and more Americans flocked to urban areas. Old methods of waste treatment and disposal were no longer satisfactory, particularly as scientists learned more about the dangers of bacteria. Students choosing the sanitary engineering option substituted classes in biology and chemistry for the classes in railway and highway engineering normally taken during their junior and senior years. The sanitary option also demanded detailed studies in hydraulic engineering. Supervising the sanitary engineering curriculum was Elton D. Walker, a graduate of M.I.T. and an engineer of wide practical experience. In 1906 the department reorganized these options into four-year courses in order to permit students to take more subjects specifically related to their interests, and Penn State became one of the first colleges in the United States to offer a Bachelor of Science degree in the field of sanitary engineering.

The fourth component of the School of Engineering, the Department of Mining Engineering, also underwent substantial improvement. Magnus Ihlseng resigned in frustration soon after the trustees eliminated the School of Mines. Marshman E. Wadsworth, former president of the Michigan College of Mines, succeeded Ihlseng as department head and immediately began strengthening the mining engineering curriculum so that one day it could resume its place as the core of an independent mining school. Wadsworth knew he faced an uphill battle. In 1901 the department's enrollment positioned it a lowly nineteenth nationally. The department possessed no adequate facilities for research and experimentation, with the exception of a fine mineralogical laboratory. As something of an unwanted stepsister of the other engineering departments, it suffered to a much greater extent than the others from a lack of space. Wadsworth blamed the General Assembly for most of the department's troubles. In his report to President Atherton in 1905, he angrily denounced the parsimony of the lawmakers in

Mechanical engineering practicum, 1903. Seated in front row, second from right, is Charles Kinsloe, who later served for nearly forty years as head of the Department of Electrical Engineering. (Penn State Collection)

Harrisburg. "The present condition is totally unworthy of such a state which stands in the front rank of mining," Wadsworth declared. "Cannot adequate support be given to enable this department, which was established by the state, to become a creditable child of its parent? Is there any reason why, except from a penurious policy, the magnificent and enormous mineral industry of the state should be so poorly recognized in the educational field?" Under Professor Wadsworth's energetic direction, the mining department ultimately experienced a renaissance, even in the absence of financial sustenance from the Commonwealth. Six years after he had taken over, undergraduate enrollment had climbed to 98, making the department the fifth largest in the nation and the largest in Pennsylvania. The department moved into new quarters in the old Mechanic Arts Building, which was enlarged to make room for assaying, metallurgical, and mining laboratories. In recognition of this progress, the board of trustees in 1906 approved the formation of a School of Mines and Metallurgy with Wadsworth as dean. Henceforth, the Department of Mining Engineering ceased to be a unit of the School of Engineering.

Whether or not they had forewarning of the overcrowded classrooms, inadequate laboratories, and overworked teachers that awaited them, students continued to enter the School of Engineering in record numbers during the early 1900s. By 1906–07, 490 undergraduates, or two-thirds of the College's student body, were enrolled in engineering. In contrast to the pre-Atherton years when many Penn State students came from Centre and adjoining counties, students now hailed from all sections of the Commonwealth. The University of Pennsylvania, Lehigh University, the Western University of Pennsylvania, Lafayette College, the new Carnegie Institute of Technology, and several smaller schools in Pennsylvania had by then established one or more engineering degree programs; but none could offer instruction at such low cost as Penn State, which took seriously the Morrill Act's charge to educate the "industrial classes" and still had not levied a tuition fee for state residents. Expenses for room and board, laboratories, and incidentals were lower than those of most competing institutions, so that a student could live comfortably spending no more than $250 per academic year.[9]

Numerous factors besides the low cost convinced students to begin their engineering careers at Penn State, but a low standard of admission was not one of them. On the contrary, prospective students and their parents often claimed that the College's entrance requirements were too stringent. To be admitted to the School of Engineering, a student had to present evidence that he possessed a solid background in such fundamental subjects as algebra, plane and solid geometry, and physics, as well as in English and English literature. He could do this in either of two ways. According to the catalog, an applicant would be accepted if he were a graduate of one of "a select list of high schools and academies in Pennsylvania whose standard of requirements has been ascertained to be satisfactory" and if he had done satisfactorily in the necessary subjects. If he did not graduate from one of these approved schools, the applicant had to pass a written examination, administered by the College, in the required subjects. In any case, admission criteria were exacting, as more than a few disappointed applicants discovered.

What did attract students to the School of Engineering at Penn State was the combination of an inexpensive technical education of superior quality and a virtually insatiable job market for engineers. No better proof exists of the excellent training engineering students received—despite the myriad of handicaps confronting the College—than the high regard which potential employers had for them. Letters from employers' representatives to the engineering faculty in 1902, for example, typically revealed these kinds of comments:

> We should very much like to get a good representation from your class in electrical engineering this year, and we will agree to place in our testing department at least six of your men. . . .

My object in writing is to find out if any of the 1902 boys would like to enter signal work. . . .

We can use two of the 1902 graduates, and I know they will get a square deal with this company. . . .

More State men are wanted here and right now![10]

The enthusiasm of employers for his school's alumni led John Price Jackson to remark in 1906 that all the engineering departments could triple their number of graduates and still not satisfy the demand. Among businesses offering the greatest number of employment opportunities for Penn State engineers during these years were the General Electric Company, Westinghouse Electric and Manufacturing Company, the Union Switch and Signal Company, and the Pennsylvania Railroad. While most alumni remained in the Northeast, a fair representation could be found nationwide and even overseas. So many civil engineering graduates were assisting in the construction of the Panama Canal, for instance, that a chapter of the alumni association was begun in the Canal Zone.

Turn-of-the-century alumni continued to bring distinction to themselves and their alma mater. Addams S. McAllister, for example, graduated in 1898 with a bachelor's degree in electrical engineering. After serving what had almost come to be a required apprenticeship for Penn State engineers at Westinghouse, he began graduate studies at Cornell and in 1905 became the first Penn State-trained engineer to earn a doctorate. He went on to achieve national recognition as a writer on engineering and business topics, working for many years as an editor at *Electrical World,* a leading trade periodical. McAllister regularly returned to Penn State as a "professorial lecturer" or adjunct professor in the School of Engineering. Charles E. Denney ('00, mechanic arts), Clarence G. Stoll ('03, electrical engineering), and Bayard D. Kunkle ('07, electrical engineering) used their technical education to good advantage in the world of business. Denney worked for nearly two decades as a signal engineer before joining the New York, Chicago, and St. Louis Railroad (Nickel Plate) as a vice-president. In 1929 he was named president of the Erie Railroad and ten years later became president of the Northern Pacific Railroad. He was by far the most famous alumnus in the history of the mechanic arts program. Stoll spent his entire career with the Western Electric Corporation, including a long stint as manager of its noted Hawthorne Works, and retired as the firm's chief executive officer. Kunkle was associated for most of his career with the General Motors Corporation, where he rose to become president of the Delco Products subsidiary and later vice-president in charge of GM's Canadian and overseas operations. Miles I. Killmer ('06, civil engineering) was perhaps Penn State's most well-known civil engineering graduate of that era. He helped design, construct, or otherwise had a hand in almost every major New York City tunnel built in the first half of the twentieth century, including the Pennsylvania Railroad's Hudson and East River tunnels, the Holland tunnels, and much of the New York subway system. Earl B. Norris ('04, mechanical engineering) and George C. Shaad ('00, electrical engineering) were only two of many engineering alumni who made education their lifelong careers. Norris served as Dean of Engineering first at the University of Montana and then at the Virginia Polytechnic Institute until his retirement in 1952. Shaad taught for many years at the University of Kansas and in 1927 became dean of its School of Engineering and Architecture. While the accomplishments of engineering alumni were due more to the almost unlimited opportunities for advancement in the technical fields than to the fact that these persons held degrees from Penn State, this circumstance cannot diminish the importance of the College as a major source of academically trained engineers.

Louis Reber knew that even in the midst of an expanding job market, the School of Engineering could not continue to produce competent graduates unless its resources were markedly supplemented. In his annual report for 1906–07, he stated that when the School had been formed in 1895, it

> was well equipped for the work it was called upon to do. Since that time, similar institutions have made large extensions to keep pace with greater numbers of students and increasing demands for specialization work. The Pennsylvania State College has experienced the same needs but has been able to add little to her resources. The time has come when expansion in all departments, in buildings, equipment, and teaching force, must be provided for, or the reputation so splendidly established must show decline.

Reber compared Penn State to the University of Wisconsin, another land-grant institution. Wisconsin, he reported, had been spending $15,000 *annually* for the last five years just on equipment for its engineering departments. The Pennsylvania State College had spent a *total* of only $10,000 on similar purchases over the last *ten years!*

In 1906 Dean Reber began work on a plan for reorganizing his school along topical rather than the traditional departmental lines. He believed this to be the vital first step in enabling the College to retain its prominence as a technical institution. Sharing a belief widely held among engineering educators of the day, Reber maintained that students needed more specialized instruction to help them meet the challenges that would confront them after they had left school and entered the real world of engineering. The old-style options would no longer suffice. He proposed to eliminate all the old departments and to create nearly a dozen new ones according to their topical function. Under this scheme, the School of Engineering would consist of these departments:

Heat Engineering
Machine Design
Railway Mechanical Engineering

Electrical Engineering
Electrochemical Engineering
Telephone Engineering

Railway Topographical and Structural Engineering
Hydraulic and Sanitary Engineering

Mechanics and Materials of Construction
Drawing and Descriptive Geometry
Experimental Engineering

The last three of these departments were to be "service" departments only, providing instruction and facilities for other departments but having no students of their own. The others would make available to their students a number of options for specialized training. Options in the railway mechanical engineering course, for example, would consist of locomotive design, car lighting and heating, and signaling.

Naturally Reber predicated his reorganization on the construction of additional buildings and an increase in the size of the faculty. Specifically, he recommended the erection of a railway mechanical engineering building, a locomotive test plant, an electrical engineering building, and a single extension to the Main Engineering Building. More teachers, he contended, could be recruited simply by raising salaries to a competitive level, although the College would have to be careful to select men who had extensive experience in a specific field. Generalists were no longer desired. The dean estimated the total cost of the reorganization, including buildings, equipment, and salaries for two years, to be $337,000. Unfortunately, President Atherton never had a chance to review Reber's proposals. After suffering ill health for many months, he died in July 1906. The trustees were still searching for his successor when Reber submitted his completed plan in the spring of 1907.

To George Washington Atherton must go the credit for a multitude of improvements at The Pennsylvania State College, which had the ultimate effect of transforming it from an educational nonentity to a respectable institution of higher learning. Nowhere were these improvements more evident than in engineering. Atherton had a clear understanding of the College's obligations under the Morrill Act and consequently did everything within his power to foster the growth of engineering education. His relationship with Louis Reber and the other engineering faculty was a model of administrative cooperation and mutual support. Without the continued encouragement of President Atherton, the School of Engineering surely would not have experienced such rapid growth so soon after its founding. So firm a friend of engineering had Atherton been that Louis Reber looked to the future with considerable anxiety. "I became really apprehensive as to what the future might bring," he later confessed. "If there was to be another Shortlidge fiasco, it might be serious for the future of the institution."[11]

In the summer of 1907, Charles Van Hise, president of the University of Wisconsin, aware of Reber's accomplishments and suspecting that he might be ready for a change, offered him the position of Director of University Extension at Wisconsin. Reber had always been a faithful supporter of extension education, although he had never succeeded in convincing Penn State's trustees or the legislature to appropriate money for extension work in engineering. When Van Hise promised him a generous increase in salary and a free hand in directing extension activities, the dean finally accepted. It was a decision not easily arrived at. Reber had deep personal and professional roots at The Pennsylvania State College. Even more than President Atherton, he was responsible for the solid foundation and steady growth of engineering education at Penn State. A superb administrator, he had run the School of Engineering for over a decade on a shoestring budget and still managed to provide students with a caliber of instruction that was the envy of many other schools. A gifted teacher, he won the unending admiration of many of these students. He had the wisdom to recognize the changes that engineering education must undergo to meet the changing demands of society, yet he tempered his visions of the future with a strong dose of realism and never asked the impossible of President Atherton, his colleagues, or his students.

Nor were Reber's professional activities restricted to the narrow boundaries of the academic world. Governor Pattison appointed him

Pennsylvania's Commissioner to the Paris Exposition of 1889 and Commissioner in Charge of Mining and Manufactures at the Chicago World's Fair in 1893. A later governor, Samuel W. Pennypacker, selected him for a similar position in 1904 at the St. Louis Exposition. On a local level, Dean Reber took a keen interest in the development of State College, the residential and commercial community that had grown up around Penn State. (It incorporated itself as the Borough of State College in 1896.) With John Price Jackson and two other alumni, James L. Hamill ('80) and Ellis Orvis ('76), he purchased a dilapidated old hotel at the corner of South Allen Street and College Avenue (where the Hotel State College now stands) and completely renovated it to provide the finest accommodations then available to visitors to the College and community. In addition, Reber played a key role in founding the borough's first electric light company and first bank, and for a few years was co-owner of the town's first newspaper, the *State College Times,* forerunner of today's *Centre Daily Times.*

Perhaps Louis Reber may be criticized for leaving the School of Engineering just at the time he feared what the future held in store for it. Did he not under such circumstances have an obligation to remain and work to ensure that the quality of instruction did not decline? On the other hand, the preceding half-dozen years had been especially frustrating for him, in light of the meager financial support accorded the College. He did far more than duty required in trying to stretch the School's miniscule budget to satisfy the needs of a growing number of students. And in the summer of 1907, his urgent plea for more aid to reorganize the School seemed to be meeting with the same negative response as had similar pleas in years past. Moreover, as Dean, he was receiving a salary by no means commensurate with his responsibilities. The time had come when he had to look to his own affairs and that of his family. Louis Reber had more than discharged his obligation to the College. Indeed, it was the College that was in his debt. The Rebers departed for Wisconsin late in November 1907. On Thanksgiving Day they stopped briefly in Chicago to have dinner with some acquaintances, Dr. and Mrs. Edwin Erle Sparks. "Little did we realize then," Louis later remembered, "that we were dining with Penn State's next president."[12]

President Sparks and Dean Jackson

When Louis Reber discovered that Edwin E. Sparks would become the eighth president of The Pennsylvania State College, he surely discarded much of his apprehension concerning the future of engineering

Edwin E. Sparks (Penn State Collection)

at his alma mater. After graduating from the Ohio State University and serving for several years as a public school teacher in the state of Ohio, Sparks had come to Penn State in 1890 to assume the position of principal of the preparatory department. This department, distinct from the collegiate departments, offered instruction comparable to the junior and senior years of high school and was designed primarily to prepare students for enrollment in the College's four-year programs. Five years later he left to attend the University of Chicago, where he earned a Ph.D. in history and received an appointment as instructor in that field. By the time he took over the presidency of Penn State, Sparks had achieved national renown as a writer, teacher, and lecturer in American history. As the new president was to show in the years ahead, his commitment to the concept of the land-grant education in general and engineering education in particular was equal to George Atherton's. In fact, because of his interest in extension education, Sparks secured for the College a larger role as a source of engineering talent and know-how than President Atherton had ever imagined.

John Price Jackson's selection as the next dean of the School of Engineering further assured the preservation of a sound engineering curriculum. The youthful head of the Department of Electrical Engineering was an obvious choice for the trustees. He had already served as acting dean while Reber was an Exposition commissioner and had worked closely with his brother-in-law in devising plans for the School's reorganization. When Professor Jackson became dean in the autumn of 1907, the School of Engineering ranked fourth nationally in total undergraduate enrollment, second in electrical engineering, fifth in mechanical engineering, and ninth in civil engineering. It was also the Commonwealth's largest engineering institution, the top five at this time being Penn State (489), Lehigh (487), the University of Pennsylvania (477), Lafayette (159), and the Western University of Pennsylvania (107).[13] Jackson and Reber agreed, however, that unless the School could make some major physical enlargements and some substantial additions to the curriculum, it would not enjoy so high a ranking for very long.

The board of trustees realized this, too. At their June 1907 meeting, the trustees responded to the reorganization proposals Reber had submitted by approving the formation of two new departments, the Department of Engineering Mechanics and Materials and the Department of Engineering Drawing (also known as the Department of Drawing and Descriptive Geometry). Charles E. Paul was appointed head of Engineering Mechanics and Materials. This department had charge of instruction in elementary mechanics and engineering materials, subjects formerly taught by the Department of Mechanical Engineering, and in applied mechanics, formerly taught under the auspices of the Department of Civil Engineering. The mechanics and materials department also took over supervision of civil engineering's materials testing laboratory. In line with Reber's recommendation, it was essentially a service department, providing training in fundamentals for students of other departments. It had no degree program of its own. When Professor Paul resigned in 1908, Paul B. Breneman was selected to head the department. After completing both undergraduate and graduate work in civil engineering at Penn State, Breneman had worked briefly for several large mining firms in the East and Midwest. His return to Penn State was a permanent one, for he continued to head the Department of Engineering Mechanics and Materials until his retirement in 1938.

The Department of Engineering Drawing initially was a service department, too. Under its first head, James B. Whitmore, it provided instruction in mechanical and geometric drawing to students in engineering, mining, and science courses; but President Sparks and Dean Jackson had more important plans for this department. In his annual report for 1908–09, Sparks made quite plain his hope that the Department of Engineering Drawing would soon become a baccalaureate curriculum in its own right. Jackson proposed that it become a department of architectural engineering, since the architectural courses of most institutions did not give adequate attention to the technical aspects of building construction. Furthermore, much of the essential subject matter of architectural engineering—mechanical and geometric drawing, engineering materials, applied mechanics, masonry, heating and ventilation, and electrical applications—was already being taught by the School of Engineering. Therefore a new course in architectural engineering would require little in the way of extra classrooms, equipment, faculty, or anything else requiring a substantial financial outlay. Impressed by the economies of the arrangement, the trustees in June 1910 approved the establishment of a degree program in architectural engineering to begin in the fall. Its objective, as expressed in the College catalog that year, was "to furnish the student with a broad and liberal training in both the aesthetic and construction sides of architecture."

A second new baccalaureate curriculum had its origin outside the School of Engineering. At a meeting in Philadelphia in 1907 with General Beaver, president of Penn State's board of trustees, Frederick Taylor, famous even then for his pioneer studies in scientific management, learned of the College's need for a successor to Reber as head of the Department of Mechanical Engineering. Taylor believed that the mechanical engineering course should include more instruction in the manufacturing processes and less in power plants and higher mathematics and recommended Hugo Diemer for the vacant position. Diemer, a personal acquaintance of Taylor's, had introduced the engineering curriculum at the University of Kansas in 1902. Largely on the strength of Taylor's endorsement, he was named head of mechanical engineering at Penn State, where he immediately began to explore the possibility of initiating a degree program aimed at training students to be managers as well as engineers. Dean Jackson enthusiastically backed Diemer's proposals and submitted a tentative outline of his curriculum to the board of trustees along with an appeal that swift and positive action be taken. "The fact that industrial education for engineers has heretofore confined itself, to quite a large extent, to the more technical side of industries at the expense of the more strictly business departments, is just beginning to be realized," Jackson told the trustees. "It is therefore desirable to add a new course, especially designed to fit men for the business positions of industrial establishment, including transportation and buying and selling."[14] American manufacturers were calling for more persons who could superintend the non-technical facets of business and industry, ranging from the manufacturing process itself to labor utilization to the logistics of transportation. The highly specialized training received by most engineers did not equip them for such a broad spectrum of duties, and curriculums in business administration were as yet far in the future. Hence, in addition to engineering fundamentals, Diemer's curriculum included such subjects as industrial organization, principles of engineering economics, advertising and salesmanship, plant layout, time and motion studies, and corporate accounting, all directed toward the goal of making the process of production and distribution more efficient. Dexter Kimball had recently begun teaching "works management" to engineering students at Cornell University, but his course was an adjunct to other engineering courses and was not as comprehensive as the one suggested by Professor Diemer.[15]

The trustees approved a two-year course (limited to juniors and seniors) in "industrial engineering" to begin in the fall of 1908 in the Department of Mechanical Engineering. Enough student interest was shown to begin a separate four-year curriculum in 1909. Diemer then resigned his mechanical engineering position to become head of the Department of Industrial Engineering—the first department of its kind in the nation. He told Dr. Sparks he expected the program to produce "men who are thoroughly familiar with the productive processes, with broad interests, and who are at the same time thorough accountants and businessmen." To facilitate instruction in the new course and to encourage the initiation of similar courses at other institutions, Diemer authored *Factory Organization and Management*, the first industrial engineering textbook and for many years the standard work in its field.

Hugo Diemer was a graduate of Ohio State University and had joined the Penn State faculty in 1907. During his earlier years as a teacher at the University of Kansas and Michigan State College and as a consulting production engineer, he had been a forceful exponent not only of managerial training for engineers, but also of vocational manual training at the secondary and post-secondary levels. He was one of the first educators to propose that technical training involving manual skills be offered in high schools and urged the establishment of institutions comparable to today's vocational-technical schools. Diemer recognized that The Pennsylvania State College, with its long-standing two-year course in mechanic arts and its well-equipped shops, was in an ideal position to produce teachers for manual training courses. With Dean Jackson's blessing, the Department of Industrial Engineering took over supervision of the mechanic arts course, which was reorganized into two-year and four-year programs and renamed the industrial education course. Both programs were geared to prepare persons to teach mechanical and industrial arts in high schools, vocational schools, and colleges. According to the College catalog, "persons of mature years with trade experience" could apply for entry into the two-year curriculum. All others had to enroll in the four-year program. During their senior year, all students in industrial education had the opportunity to do "practice teaching" in central Pennsylvania high schools.

The creation of the several new departments intensified the School of Engineering's need for more space. To relieve some of the pressures caused by overcrowding, plans were drawn up to convert some of the rooms in the basement of Old Main to electrical engineering laboratories—an unmistakable sign of the "retrogression" Louis Reber had warned of earlier. Fortunately, General Beaver intervened and personally solicited enough funds from friends of the College to pay for the erection of a "temporary" electrical engineering building. This two-story, wood-frame structure, built in 1908, stood just to the rear of the President's House and measured 150' × 40'. Originally denominated the Engineering Extension Building, it was soon rechristened the Engineering Annex in order to avoid confusion with the School of Engineering's newly organized extension activities. Eventually it came to be known as Engineering Unit F. On the ground floor the building housed electrical and electrochemical labs; the second floor contained miscellaneous classrooms and the office of the Department of Engineering Drawing (soon to become the Department of Architectural Engineering). Dean Jackson regarded the building as a stopgap measure rather than as a permanent solution. After underscoring the fire hazard inherent in placing electrical laboratories in a wooden structure, Jackson in his annual report for 1909–10 requested nearly $700,000 for four new buildings: a mechanic arts and shops building, a transportation building, an electrical engineering building, and a new hydraulic laboratory building (to be located at nearby Thompson's Spring). President Sparks, possibly a bit more realistic in his appraisal of the attitude of the Commonwealth's lawmakers, asked for only $500,000 for new engineering buildings that year. In any case, the legislature approved nothing even close to either amount. It allotted a mere $20,000 for one new building and no money at all for the equipment to be installed within it. Jackson then revised his building plans and in 1911 submitted a much less expensive version than that of the previous year. The new scheme still called for four additional structures, including the one for which the Commonwealth had already allotted money; but they were to be much smaller and less elaborate than he would have preferred. All would be characterized by a maximum of space and a minimum cost of construction, and all were to be

Engineering Unit F. In this 1930s view, a portion of the thermal laboratories can be seen at extreme right. (Penn State Collection)

situated to the rear of the Main Engineering Building. Two of the buildings would provide quarters for the Department of Electrical Engineering, while mechanic arts shops and mechanical engineering laboratories were to be located in the other pair. Jackson also submitted a less detailed second phase of this master building plan, consisting of two transportation units and a new power plant.

Meanwhile construction of the building authorized by the General Assembly proceeded swiftly throughout 1911. This structure, soon to become known as Engineering Unit E, according to President Sparks "was intended to exemplify the ideal engineering building for college instruction and experimental use" and was expected to set the pattern for the other units.[16] A stark, utilitarian three-story building composed of red brick, it was dedicated in November 1911, with Governor John K. Tener on hand for the ceremonies. The Department of Electrical Engineering, which still had by far the largest student enrollment, occupied most of the space in the new unit. Governor Tener, unlike his immediate predecessors, proved to be a friend of Penn State and of higher education in general. At his insistence, the legislature in 1913 allocated $400,000 in construction funds to the College, to be used as the trustees and President Sparks saw fit. This was the first instance of such a general appropriation, since on all previous occasions the legislators had designated a specific amount for a particular building. After studying a number of proposals, Sparks and the trustees decided that the needs of other schools were more urgent than those of the School of Engineering. Engineering subsequently received only $25,000, or enough to build a companion unit to the one just completed. Most of the $400,000 appropriation was used for the construction of a new mining building, a horticulture building, a chemistry laboratory, the first wing of a new liberal arts building, and a livestock judging pavilion. The second engineering unit (later known as Unit D) was erected immediately adjacent to the first and was its virtual twin, architecturally speaking. The remaining electrical engineering laboratories were moved from the Annex (Unit F) into the new structure, their place being given over to rooms for architectural engineering and storage space for the Department of Civil Engineering.

More buildings were of little value in themselves unless additional laboratory apparatus could be procured. Dean Jackson warned in 1911 that the lack of equipment seriously undermined the quality of instruction. "We have not kept abreast of the advance in the mechanic arts or of the increase of our own teaching duty, nor have we kept abreast of the most efficient of the richer technical colleges," he declared in his annual report. "In a surprisingly large number of instances, at the present time, students are delayed in performing their scheduled duties by reason of lack of sufficient apparatus to go around, and even more often they are compelled to use instruments unfitted for the particular work in hand."

This did not mean that the School had failed to acquire any significant laboratory equipment during the last several years. In 1906 the Pennsylvania Railroad loaned to the College an old but still operable steam locomotive (No. 01001, a Class D8b 4-4-0), but to take maximum advantage of this valuable acquisition, a dynamometer car was needed for use in making various calibrations. As no such car could be secured at an affordable price, the mechanical engineering students built one of their own, capable of measuring drawbar pulls up to 15,000 lbs. A typical day of testing saw students run the locomotive and dynamometer car over the tracks of the Bellefonte Central to Bellefonte. There a short string of freight cars was coupled behind the dynamometer for the return upgrade to State College. During this portion of the trip students collected data on the amount of coal burned, steam consumed, and water evaporated in order to obtain an accurate picture of overall locomotive performance. The quality of the School's railway mechanical engineering studies must have impressed the PRR, for the railroad donated one of its own dynamometer cars (of 28,000 lb. measurement capacity) to the College in 1911. Along with the locomotive, a Westinghouse air brake rack, and an automatic block signal installation, the new dynamometer car gave the School of Engineering the necessary facilities for training railway mechanical engineers and enabled it to introduce a four-year course in railway mechanical engineering in 1911, under the direction of A.J. Wood. The absence of a locomotive test plant, nevertheless, prevented the railway mechanical engineering program from realizing its full potential. In fact, the Pennsylvania Railroad had given the College the locomotive on the condition that the School of Engineering would eventually build an indoor test plant on which it could be used, which explained why Dean Jackson had continually stressed the need for a fully equipped transportation building. His anxiety over losing the engine when no money could be secured for such a facility dissipated in 1912, however, when the railroad made an outright gift of the locomotive and its dynamometer car.

An equally unique addition to the laboratory resources of the Department of Mechanical Engineering was the "aeroplane test track," constructed in the spring of 1911. A small car propelled by a 20-horsepower gasoline motor circled a track 200 feet in diameter at speeds up to 48 miles per hour. It allowed students to test the lifting effects of planes, as well as to determine the effects of wind resistance on various configurations of automobile fronts.

The apparatus that aroused the most popular interest belonged to

Mechanical engineering seniors and faculty with the College locomotive and dynamometer car. (Penn State Collection)

the Department of Electrical Engineering. In the fall of 1910 students in that department installed a wireless telegraph station in Unit F. Using a temporary tower that had been erected atop the building, student operators were able to communicate with other wireless stations within a 200-mile radius and often with coastal stations and ocean vessels. In October 1911 results of the Penn State-University of Pennsylvania football game were relayed back to the College at regular intervals from the Marconi station on the roof of Wanamaker's Department Store in Philadelphia, marking the very first "broadcast" of a Penn State gridiron contest.[17] The most significant addition to the Department of Civil Engineering's physical facilities was a 50-foot section of concrete highway laid down in 1911 under the supervision of the faculty. The experimental pavement was to be studied for its durability and was an important resource for the new course in highway engineering, the third (after civil and sanitary engineering) degree curriculum in the civil engineering department.

The Department of Mechanical Engineering also added a new degree program at about this same time. The Pennsylvania State Millers Association in 1910 requested the College to initiate a course for the training of flour mill engineers. The trustees, President Sparks, and Dean Jackson were only too happy to comply, since the Millers were willing to underwrite most of the extra costs incurred. Benjamin W. Dedrick, organizer and first president of the Technical Millers of the United States, was offered the opportunity to supervise the new course. A practicing miller from the Midwest, Dedrick hesitated to accept the appointment. He worked in a field that employed few if any academically trained engineers and admitted later that "I always had the idea that a college man was a snob and that I would find it hard to get along in these new surroundings."[18] But at length he agreed to take on the assignment, which consisted of nothing less than launching the nation's first baccalaureate program in milling engineering. Dedrick's early misgivings soon evaporated as he received "perfect cooperation" from Jackson and Sparks in converting the south end of Unit F into a temporary mill, replete with milling machinery, granulator, sifter, scales, dust collector, and other equipment that could be used to instruct students in proper mill design and operation. Dedrick was also one of the few members of the engineering faculty prepared to allot much time for research. In 1912 he won a modest contract from the United States Department of Agriculture to devise methods of preventing dust explosions in flour mills, the chief safety hazard in the milling industry. The grant constituted the first federal aid received by the School of Engineering in support of research activities.

An asset equal in value to any laboratory apparatus was the library. In the 1890s, the College library was housed in Old Main, while each engineering department maintained its own "reading room" containing material of special interest to engineers. Around 1900 the need for more classroom space in the engineering buildings forced the consolidation of the reading rooms into a single small room in the Main Engineering Building. Shortly after the new Carnegie Library was completed in 1904 (financed by a $150,000 gift from trustee Andrew Carnegie), most of the reading room's bound periodicals, reference works, and text books were transferred there to form the core of a 1500-volume "engineering alcove." In Carnegie Library, the engineering works came under the administration of professional librarians and were made available to students for longer and more convenient hours.

Beginnings of Research and Extension Education

Two events occurred during John Price Jackson's tenure as dean that symbolized more than any others the maturation of the School of Engineering. One was the establishment of an engineering experimentation station, the idea for which can be traced back at least to 1902, when Louis Reber brought the question before the board of trustees. But to Dean Jackson must go most of the credit, for he turned the recommendation into reality. Both Reber and Jackson were sensitive to the need to use the theoretical knowledge gained in the classroom and laboratory to solve the practical problems of industry. An experiment station would provide an ideal environment for faculty and advanced students to pursue research that had important applications. Penn State's agricultural experiment station, in operation since 1887, had proved highly successful in this respect, as had those at many other land-grant institutions. Whereas the federal government provided direct assistance to agricultural experiment stations through the Hatch Act of 1887, it refused to support research in engineering, the rationale being that agricultural research benefitted a broad section of the population but advances stemming from engineering research tended to be advantageous to a narrow sector of industry with only indirect benefits to the general public. Nor was the Association of American Agricultural Colleges and Experiment Stations—the collective body of land-grant schools—much interested in engineering; thus it did not press Congress to enact legislation providing money for engineering experiment stations. The Association preferred to focus on agricultural education and research and between 1903 and 1919 did not even have an

John Price Jackson (Penn State Collection)

engineering section in its organization. Political lobbying and indeed the entire responsibility for shaping the development of engineering education was left to the Society for the Promotion of Engineering Education (founded at Chicago in 1893) and the several professional societies, principally the American Society of Civil Engineers, the American Society of Mechanical Engineers, and the American Institute of Electrical Engineers. In 1912 the Land-Grant College Engineering Association was formed, and it spent seven years struggling in vain for Congressional passage of an engineering version of the Hatch Act before finally being reconstituted as the engineering section of the American Land-Grant College Association. Given the unfavorable climate, only two land-grant schools, the University of Illinois in 1903 and Iowa State College in 1904, had been able to convince state legislatures or private donors to underwrite the cost of an engineering experiment station.[19] Edwin Sparks wholeheartedly endorsed the concept of an engineering experiment station for Penn State, concluding that if the achievements of its agricultural counterpart could be duplicated, the station would make many corporate and individual friends for the College throughout the Commonwealth. It would be especially useful for governmental bodies and small firms whose resources did not permit them to conduct technical research on their own. The trustees, too, were eventually convinced of the soundness of the idea, and in May 1909 they approved the creation of an experiment station within the School of Engineering. The station's goals, as outlined in its first *Bulletin,* were

> to carry on scientific investigation concerning the problems of engineering and the mechanic arts; to study and report on engineering methods, materials, and processes relating to manufacturing, transportation, sanitation and water supply, structures, and the various other industrial interests of Pennsylvania; to furnish specific instructions which may prove of use to the masses of the people.

Unfortunately, no significant amount of money accompanied the establishment of the station. The College could not afford to endow it and was unable to persuade the Commonwealth to do so. The station had neither staff nor laboratories that it could call its own, nor did it even have a fulltime director. Faculty members who conducted research at the station did so in addition to fulfilling their obligations in the classroom. What limited funds that were available to support research came from the School's general budget. Results of the experiments were published in the *Engineering Experiment Station Bulletin,* which first appeared in June 1910. That issue contained Associate Professor of Electrical Engineering Charles L. Kinsloe's "Results of Experiments on the Effects of the Form of A.C. Waves on the Life and Efficiency of Incandescent Lamps," and Professor of Civil Engineering Elton D. Walker's "Practical Suggestions for the Construction of Concrete Floors."

A most welcome upgrading of the facilities of the station came in June 1911 with the completion of a thermal laboratories building, located between the Bellefonte Central station and Unit F. Designed principally by Professor of Mechanical Engineering Louis A. Harding, the cork-lined laboratories could maintain constant temperatures over long periods of time, thus providing the conditions required to study the properties and effects of heat transmission, refrigeration, and insulation. Soon after the thermal laboratories were put into operation, the Pennsylvania Department of Highways awarded its first research contract to the College: a grant to study the effects of temperature extremes on various types of pavement.

The second event that marked the School of Engineering's coming of age was the beginning of extension education in engineering. Engineering extension shared the same general goal as the experiment station, that is, to put the human and material resources of the School of Engineering to work for the citizens and industries of the Commonwealth, in a much more direct and practical manner than heretofore had been possible. Dean Jackson believed that engineering education could be of great benefit to far more people than the relative handful who were able to spend four years at the College. As early as 1903 he noted that "there is a greater and greater demand for practical instruction to artisans," yet few of them had the time, money, or academic preparation to work for a college degree. Therefore, Jackson had urged that his electrical engineering department be allowed to offer "elementary courses in electrical engineering in the classrooms, laboratories, and drafting rooms for foremen, wiremen, and other artisans." These would be short courses, lasting from a few days to a few months, and would focus strictly on practical applications rather than theory. For those individuals who could not come to the College even for a short while, Jackson proposed correspondence courses, in spite of the fact that he considered it "manifestly impossible to obtain as satisfactory results by this method."[20] Fiscal restraints and the physical limitations of the School of Engineering prevented Jackson's suggestions from being adopted in 1903. When he became dean, Jackson renewed his campaign for an engineering extension program, predicting that it "can eventually be of great service to a large industrial population that is not now reached by any other fully satisfactory educational influence."[21] As a first step in advancing engineering education beyond its traditional academic setting, Jackson encouraged members of the faculty to tour the state giving illustrated lectures on engineering topics of special interest to selected groups. Professor of Mechanical Engineering Arthur H. Gill, for example, on several occasions in 1909 spoke before various industrial organizations in the Harrisburg and Steelton areas on fuel economy and smoke consumption.

Two other milestones in the growth of engineering extension occurred that year. First, the school board of the city of Williamsport, a lumber and railroad center 60 miles northeast of the College, invited the School of Engineering to set up an evening course in mechanical drawing. Next, at the behest of the Pennsylvania Railroad, the School organized evening courses in elementary electrical and mechanical engineering for apprentice mechanics at the road's Altoona shops. In both cases the School of Engineering supplied the instructors and planned the lessons, and the sponsoring body paid all expenses. These initial ventures in extension work were greeted with unqualified success. The Williamsport Vocational School grew rapidly, but since the level of instruction was not really at the college level, Dean Jackson arranged to have Pennsylvania's Department of Public Instruction assume supervision of the school only two years after it opened. So delighted was the PRR with its apprentice school that it asked Jackson

The first engineering extension class at the Williamsport Vocational School, 1910. (Penn State Collection)

to organize similar courses elsewhere on the system. By the end of 1913, over a thousand PRR employees were attending extension classes in Philadelphia, Pittsburgh, Harrisburg, Altoona, and Wilmington, Delaware. Meanwhile boards of education and business firms across the Commonwealth besieged the School of Engineering with requests to launch extension courses in their own communities. Soon classes in several varieties of manual trades and in elementary engineering theory were underway in Philadelphia, Chester, Allentown, Sunbury, Tyrone, Clearfield, Erie, and a half-dozen other locations. Jackson reluctantly turned down the applications of many other communities that desired extension classes, since the School's budget and manpower could be stretched only so far. True, the School did not have to bear most of the immediate expenses of these classes, but it had no money to hire full-time personnel, other than two field directors, to plan and coordinate the statewide programs. As in the case of the experiment station, the faculty who planned the lessons and occasionally taught the classes did so in addition to, rather than in place of, fulfilling their normal teaching responsibilities at the central campus. There was no money even to hire a full-time director of extension. Dean Jackson himself acted as director until 1914, when James A. Moyer of the Department of Mechanical Engineering succeeded him as acting director. Like Jackson, Professor Moyer had to pay all his own traveling and other expenses incurred as a result of extension duties. This stipulation, ludicrous as it may have been, was a necessary one, for despite the popularity of the extension classes, the legislature had not appropriated a single dollar to support them. The School of Engineering operated extension strictly on a self-sustaining basis.

Several of the faculty began preparing lessons to be used in correspondence courses, another area in which agricultural educators had achieved much success. However, Dean Jackson reported to President Sparks in 1910 that "lack of funds and a sufficient teaching staff have made it unwise to progress far along these lines." Given the already heavy teaching load shouldered by most of the faculty, and the increasing burden of extension work, he did not want to risk spreading his resources too thin. Consequently, the initiation of correspondence courses had to be delayed pending an improvement in the School's financial position. In spite of this setback, to Penn State must go the honor of being the first degree-granting institution in the United States to put engineering extension on a formal and permanent basis.[22]

While not an element of the College's extension program in the absolute sense, another event occurred in 1910 that also helped the School of Engineering to take its work directly to the residents of the Commonwealth. Late that year Dean Jackson secured the cooperation of the Pennsylvania Railroad, the Pennsylvania Department of Highways, and the United States Bureau of Roads in outfitting a "Good Roads Train." Consisting of seven cars equipped for presenting illus-

Dean Jackson's Good Roads Train, location unknown. (Penn State Collection)

trated lectures and containing scale-model displays of highway construction machinery and methods, the train embarked on a statewide tour in January 1911. Over 50,000 people attended the lectures and passed through the exhibit cars in the course of the train's eight-week journey. The Good Roads Train was primarily intended to arouse popular sentiment in favor of better highways for Pennsylvania and contributed to the passage later in 1911 of the Sproul Road Act, which established a comprehensive system of state-maintained highways. Jackson astutely reasoned that the train would at the same time serve as an exceptional public relations tool for the College and particularly for the School of Engineering. The train and its contents would demonstrate to the taxpayers the practical advantages to be derived from adequately maintaining an institution that produced capable engineers and that carried on research aimed at improving the quality of life in Pennsylvania.

The successes of the Good Roads Train and Jackson's entire campaign to make the School of Engineering more responsive to the needs of its statewide constituency undeniably reaped a rich harvest of good will for The Pennsylvania State College. On the other hand, these successes resulted in a development that was not so favorable to the College. Governor Tener, impressed with Jackson's community-minded programs, asked the Dean of Engineering to come to Harrisburg in 1913 to become the Commonwealth's first Commissioner of Labor and Industry. Jackson accepted, with the expectation that he would return to the School of Engineering after completing his term of public service. President Sparks labeled his absence for even such a short time "a deplorable loss," although he was confident that Jackson would be an outstanding public servant.[23] As events transpired, the loss was of much greater magnitude than Sparks at first contemplated. In 1915 Jackson complied with the request of Martin G. Brumbaugh, who succeeded Tener as governor, to remain in the cabinet post for another term. After taking a year off to serve as a Colonel in the United States Army during World War I, he completed his service as Commissioner of Labor and Industry in 1919. After living for six years on the small salary Pennsylvania allotted its public officials, Jackson had little desire to return to the low pay that accompanied academic life. Instead he accepted an offer to become personnel director for the Consolidated Edison Company of New York City. He retired from that position in 1938 and died ten years later, on April 1, 1948. Ironically, Jackson's death preceded by just six weeks the passing of Louis Reber, who had retired in 1926 as Dean of Extension Services at the University of Wisconsin.

While the expansion of the curriculum, the beginning of extension, and the founding of an experiment station stand as the most prominent accomplishments of John Price Jackson's tenure as dean, they should not overshadow other achievements that also contributed to the strengthening of the School of Engineering. One of the most important of these was the steady growth of graduate education. The College catalogs reveal that graduate studies leading to technical degrees were available in engineering since the creation of the Department of Civil Engineering in 1881. Few, if any, students took advantage of the opportunity to pursue graduate work until 1887–88, when two students enrolled for advanced study in electrotechnics. However, the first graduate degree in engineering was not awarded until 1892, when John Price Jackson received the degree of Mechanical Engineer. Charles E. Maxwell, of Trenton, New Jersey, in 1893 became the first student to earn the graduate degree of Electrical Engineer. (He had received a Bachelor of Science degree in that subject from Penn State two years earlier.) Graduate students were enrolled in civil engineering as early as 1889, but none completed the requirements for the degree of Civil Engineer until 1897. Paul Breneman, E.P. Butts, and E.W. Bush—all undergraduate alumni—received advanced degrees that year.

The School of Engineering had to utilize nearly all its resources just to meet the demands of increasing numbers of undergraduates, so for a number of years graduate education remained a minor function. Not until 1910 did the number of graduate students in residence for any single year exceed three. Requirements for a degree included a year of resident instruction together with one year of research leading to the preparation of a satisfactory thesis. The College waived the need for resident instruction if the student had accumulated at least three years of experience in a relevant field. Beginning about 1904, three years of work experience would suffice only if the student had received his undergraduate degree at Penn State. If not, he had to give evidence of having completed five years of related work, a change that was of little consequence, for the preponderance of graduate students were and continued to be Penn State alumni. Most students preferred to earn their degrees on the basis of experience, rather than by the more expensive method involving a year's residence at the College. Usually the conferring of advanced degrees carried more prestige than material reward, because most sectors of industry had little need for graduate engineers. Ostensibly, one of the main objectives of graduate education in most fields was to promote research skills; but with few exceptions (principally in the electrical and chemical industries) technological progress had not yet come to depend heavily upon scientific discoveries, and research and development received only minor attention. Furthermore, most engineering educators lacked the academic preparation and the experience that qualified them to direct graduate research projects. Penn State and most other engineering institutions therefore awarded technical degrees, recognizing superior applications of existing knowledge, rather than the research-oriented degrees of Master of Science and Doctor of Philosophy.

Penn State's School of Engineering was also typical of most of its counterparts at other colleges and universities in that it had encouraged the trend toward undergraduate specialization. This trend disturbed many educators, who even before the turn of the century contended that increasing amounts of time spent in specialized studies came at the expense of general education requirements. As a result, engineering schools were granting degrees to persons who were expert in narrow technical fields but whose store of general knowledge was woefully limited. The mounting intensity of this criticism prompted the Society for the Promotion of Engineering Education to join with the national professional engineering societies (the American Society of Civil Engineers, the American Society of Mechanical Engineers, the American Institute of Electrical Engineers, and the American Chemical Society) in appointing a "Joint Committee on Engineering Education" to investigate the problem of overspecialization. Funded by the Carnegie Foundation for the Advancement of Teaching and supervised by Professor Charles R. Mann of the University of Chicago, the inquiry dragged on for many years. The Committee's report, finally issued in 1918, called for a return to a more uniform curriculum and a greater role for intellectual studies in engineering education.[24] John Price Jackson saw no need to await the findings of the Mann Report before implementing changes in his school's curriculum. Although he did not believe that the quest for specialization had fragmented Penn State's engineering program to the same extent as at some other institutions, he did concede that perhaps technical studies had received undue emphasis in recent years. With the encouragement of President Sparks, the dean appointed a committee to examine the advisability of adding more liberal studies to the curriculum. The committee deliberated for nearly a year, during which time it sought the views of educators from technical and non-technical fields alike. In 1911, upon recommendation of the committee, the School of Engineering broadened its non-technical requirements to encompass American industrial history, English economic history, labor problems, and similar subjects. Jackson reported with satisfaction that because of these changes, engineering students at Penn State took more credits in general education than did their counterparts at any other comparable institution. President Atherton's belief, expressed nearly two decades earlier, that the College's first duty was to produce the "all-around well-trained man" had not been forgotten.

Another modification made to the curriculum in 1911 was the elimination of the summer practicum. Students were now advised to obtain work experience during the summer months, perhaps serving as a surveyor's helper or an apprentice mechanic, and received six credits for each summer of work. If suitable employment could not be secured, students were expected to take additional subjects during the academic year in order to earn the 160 credits needed for graduation.

The learning process for student engineers did not confine itself to the classrooms and the laboratories, of course. Participation in the activities of the various student engineering societies served as a valuable source of knowledge and experience, too. The first group to be organized was the Electrical Society, formed under the sponsorship of John Price Jackson in 1894. It limited its membership to juniors and seniors. Juniors and seniors also predominated in the Civil Engineering Society, formed in 1901, although sophomores were admitted as associate members, a term no doubt used to make certain that the lower classmen maintained proper respect for their more advanced fellows. A Mechanical Engineering Society was established a few years later. All three groups held both professional and social activities. Each group convened regular meetings, at which faculty, distinguished alumni, and other prominent engineers discussed outstanding engineering issues and accomplishments of the day. Banquets and social gatherings were also scheduled periodically. As the years passed and the size of the student body grew, the societies served as a convenient way for students to establish close personal friendships with one another. Dean Jackson heartily encouraged the activities of these organizations, and at his direction, space on the first floor of the Main Engineering Building was converted to an "Engineering Club Room," where students, faculty, and guests could gather for social and professional functions.

One of the most noteworthy accomplishments of students during the Jackson years was the publication of *The Engineer,* commencing in March 1908. In its first issue this journal resolved "to foster interests centering around the School of Engineering, to stimulate new activities, and to make strong friends for the College." Initially, students and faculty collaborated on *The Engineer,* with Professor A.J. Wood acting as editor-in-chief. After a few issues, however, it became almost exclusively a student-run periodical under general faculty supervision.

It carried articles by alumni, faculty, and students, as well as news about the School of Engineering and its graduates. Unfortunately, financial problems and student editorial rivalries brought *The Engineer* to a premature end. It ceased publication with the May 1910 issue.

To be sure, the activities of the engineering students did not always meet with the approval of the dean and the faculty. A common practice among teachers at that time was to write on the blackboard the questions to an examination the evening before it was to be presented to the class. One night while Professor Louis Harding was busily transferring test questions from his notes to the board, he chanced to notice in the slit between the window sill and the drawn blinds several sets of eyes intently peering in at him. Saying nothing, Harding quickly finished his work and left the room, knowing full well that within a few hours nearly every student in his mechanical engineering class would have the questions—and soon thereafter the answers—to the following day's exam. Later that night he returned to the room, erased from the board all the questions that had been written there, and proceeded to chalk up an entirely new set of problems. Harding recalled that never in his life did he witness so much eye-popping and jaw-dropping as he did among his students the next morning as they filed into class and glanced at the blackboard.[25]

On another occasion, during the fall of 1909, a group of would-be engineers decided to conduct a first-hand inspection of the right of way of the Bellefonte Central Railroad. After "requisitioning" a hand car under the cover of darkness from a railroad maintenance shed at Strubles, the gang pumped their way to Alto, at the summit of the grade between State College and Bellefonte. From there they merrily coasted the remaining twelve miles to the county seat, only to be violently overturned at a half-opened switch and then greeted by unsmiling Bellefonte Central officials and police officers. Each student was fined five dollars and subjected to some good-natured ridicule in the 1910 *La Vie*, the College yearbook. The Bellefonte Central also figured in a student activity that gave more lasting pleasure. From the station along College Avenue, the tracks climbed a short but steep grade as they headed toward the wye at Strubles. A favorite pastime of some students, especially those in the railway mechanical engineering course, was to grease the rails on this grade. This was most easily done in the winter, just before the five o'clock train departed for Bellefonte, so that darkness hid the culprits from detection. As the train pulled from the station and began ascending the hill, the engine's driving wheels would hit the grease and spin helplessly, accompanied by the uproarious howls of onlooking students. The sky glowed red as the locomotive's carefully tended fire was sucked out through the stack in a shower of sparks and cinders, while the air around the cab was said to have turned a distinct shade of blue.[26]

Among the faculty, contributions of several of the more prominent members have already been discussed. A few administrative changes are also worth mentioning. Louis Harding, who was so instrumental in the establishment of the thermal laboratories, succeeded Hugo Diemer in 1909 as head of the Department of Mechanical Engineering. In 1912 Harding left to join the mechanical engineering department at Cornell University and went on to become a nationally known authority on heating, ventilation, and refrigeration, heading his own construction firm in Buffalo, New York, for many years. In 1958 his widow, Charlotte Hanes Harding, willed $300,000 of her estate to establish a loan fund in her husband's memory for mechanical engineering students at Penn State. Harding was in turn succeeded by James A. Moyer, who assumed the extra responsibility as acting head of extension in 1914. In the Department of Electrical Engineering, Charles L. Kinsloe was named head after John Price Jackson vacated the post to become dean. Kinsloe was to head the department for the next 37 years, until his retirement in 1945. Fred Foss headed the Department of Civil Engineering until 1907, when he resigned to join the civil engineering faculty at the Carnegie Institute of Technology. Elton D. Walker was then appointed as the new head of the department. These and other faculty members whose mention space does not permit served during an especially trying period. Due in large measure to the assistance of Governor Tener, who continually prodded the legislature to live up to its financial obligations to the College, John Price Jackson succeeded in enlarging the teaching staff from 14 to 55 during his tenure as Dean. Unfortunately, this increase barely allowed the School to keep pace with rising enrollments, as the demand for engineers continued to outstrip the supply by a wide margin. Little money was available to alleviate the problems caused by insufficient and outmoded laboratory equipment, and the faculty was hard pressed to keep instruction abreast of the latest technological developments.

3 Consolidation and Retrenchment: 1915–29

Elton Walker was named acting dean of the School of Engineering upon John Price Jackson's departure for Harrisburg late in 1913. When Jackson subsequently decided to remain in state government longer than he had originally anticipated, he submitted his resignation. President Sparks and the trustees then began searching for a permanent replacement. After having carefully scrutinized the credentials of a sizeable number of prominent engineering educators, they offered the position to Robert Lemuel Sackett, a professor of sanitary and hydraulic engineering at Purdue University. Sackett accepted at once and arrived at the Penn State campus during the summer of 1915. A native of Michigan, the 48-year-old Sackett had earned both undergraduate and graduate degrees from the University of Michigan. He worked briefly for the Corps of Engineers before joining the faculty of Earlham College in Richmond, Indiana, as an instructor in the mathematics and engineering department. He had risen to the headship of that department by 1907, when he received an appointment to the civil engineering department at Purdue. Sackett's eight years at that institution were distinguished both by his effectiveness in the classroom and his interest in practical engineering affairs. A nationally recognized expert in the field of municipal sanitation, he served as engineering advisor to three successive Indiana governors and a host of local government bodies. In characterizing Robert Sackett, his contemporaries invariably took note of his iron will and his rigid determination to do his job, regardless of the magnitude of the obstacles thrown in his path. His tenacity reflected itself in his demeanor. A former student later recalled, in a typical observation, that the new dean "was a big, heavy-set man with a growling voice who could have just scared the dickens out of most undergraduate students, and probably did."[1] Yet beneath this gruff exterior existed a warm and sympathetic human being. Sackett was a man who felt a deep personal concern for the welfare of his students. The School of Engineering was fortunate in obtaining a leader who possessed these seemingly contradictory qualities. The years ahead would so severely tax his persevering nature that lesser men would surely have thrown up their hands in despair. Students and faculty alike would face unprecedented hardships. The resolute, uncomplaining example set by their dean would make their own sacrifices and privations infinitely more tolerable.

New Leadership for the Crisis Years

At the outset of the Sackett administration, enrollment in the School of Engineering accounted for 762 of the College's 2099 undergraduates. Faculty (excluding assistants) numbered fifty-seven. Five departments offered baccalaureate degree programs in eleven specialized areas, in addition to which the Department of Engineering Mechanics and Materials operated the materials testing laboratory and gave instruction to students of many departments in the Schools of Engineering, Mines and Metallurgy, and Natural Science. The School of Engineering was housed in four major structures: the Main Engineering Building and Engineering Units F, E, and D. A few smaller buildings, the campus power plant, the thermal laboratories, and the College's new sewage treatment facility (completed in 1915, on the site of the present-day plant) completed the group of buildings that were used for instructional purposes.

The primary research arm of the School, the Engineering Experiment Station, continued to make extensive use of the thermal laboratories. Professor A.J. Wood of the mechanical engineering department, a guiding force behind the implementation of the School's railway mechanical engineering course, was now in the process of shifting his research interests to the area of heat transmission and refrigeration. At the time of Sackett's arrival, Wood was carrying on Harding's investigations into the development of more efficient insulating materials. He had general charge of all activities at the Experiment Station, although Dean Sackett was its official director. Despite the value of the work being done there, however, a shortage of funds was a constant restraint on the scope of the research.

Sackett found the situation with regard to the School's off-campus

Allentown branch school, the first permanent engineering extension center. (Penn State Collection)

affairs more satisfactory. Extension activities were now organized into a separate division and were still enjoying a period of steady growth. By the spring of 1916, over 3500 students were attending classes at thirty-eight "cooperative centers" around the state. Co-sponsors of these classes ranged from local school boards to the Young Men's Christian Association, from the Philadelphia Navy Yard to the Eastern State Penitentiary. The subjects taught by the extension faculty had progressed far beyond the rudimentary mechanic arts classes initially offered and now included such college-level topics as power plant economics, automobile engineering, shop management, and architectural drawing. In order to bring some continuity to extension instruction, the School of Engineering in cooperation with a local citizens' committee opened a "branch school" at Allentown in 1912–13. Unlike the other extension centers, the branch school was organized on a permanent basis, so that students had an opportunity to take a series of related courses over a period of years, at the end of which certificates were awarded. Convening initially during the evening hours in a former elementary school, the Allentown branch school was destined to become the first of nearly a dozen permanent sites for engineering extension education scattered across Pennsylvania. Extension Division

Supervisor James A. Moyer, who along with John Price Jackson had vigorously promoted the concept of extension education, won such acclaim for his work in this field that the state of Massachusetts asked him to become director of its entire technical extension program. (Massachusetts operated its extension service through its Department of Public Instruction.) When Moyer left in October 1915 to accept this new position, Sackett appointed Norman C. Miller, a 1912 mechanical engineering graduate of the University of Michigan and extension field director for eastern Pennsylvania, to head the Extension Division. The Dean himself took over temporary direction of the Department of Mechanical Engineering for the remainder of the academic year. In September 1916, Edward P. Fessenden, formerly of the mechanical engineering faculty at the University of Missouri, became the new department head. The advances made by the Extension Division notwithstanding, Dean Sackett indicated in 1916 in his report to President Sparks that there existed considerable room for further improvement. Extension education, he pointed out, was "being carried out by the regular staff of the School of Engineering. No extension of the work to new centers will be possible until sufficient appropriations are made to employ a staff of instructors who can devote their entire time to Extension work." Sackett noted that because of the lack of money, "It is also impossible for us to do any considerable amount of correspondence work, although many requests for the same are received." The extension staff, consisting of two field directors and one supervisor, was much too small to meet all the demands placed upon it.

Nor did the Dean find much cause for cheer in other sectors. On the positive side, "the marked spirit of loyalty to the College and to the students" that he discovered among the faculty during his first year was a source of much gratification. As for the students, Sackett was pleasantly surprised by their "high moral character and the diligence with which they pursue their studies."[2] Unfortunately, the School of Engineering had not been able to admit as many students as it wished. Ignoring the School's appeal for more room, the legislature had not granted it any building funds for the 1915–17 appropriations period. In fact, the lawmakers denied President Sparks' request for more money to support new construction of any sort. Insufficient space for classrooms and laboratories forced the School of Engineering to continue its policy, begun in 1913, of turning away significant numbers of *qualified* applicants for admission. Prior to this time, only those prospective students whose academic accomplishments had been judged unsatisfactory had been denied entry. The growth of the School, so pronounced since the turn of the century, thus ground to a halt, with total enrollment averaging between 700 and 800 students during the years immediately prior to World War I.

Sackett must have judged this situation to have bordered on the intolerable, yet he did not publicly rebuke the Commonwealth for its parsimony, nor did he give any hint that he regretted his decision to come to the College. Privately, however, he contended that to deny qualified residents of Pennsylvania the opportunity to pursue engineering studies at the College because of inadequate room was grossly unfair. As the Commonwealth's only land-grant institution, Penn State was the only school in Pennsylvania where students could receive an engineering education at a relatively low cost. Sparks and the board of trustees did not have to be convinced of the merits of Sackett's claims, of course. In the fall of 1916, as part of a requested biennial appropriation of $1.1 million, President Sparks asked for $250,000 to finance construction of two more engineering units and the purchase of a new hydraulic lab. "If this sum seems large," Sparks said in presenting his case to the General Assembly, "it must be remembered that the state is far in arrears in meeting the demand."[3] The legislature approved only $100,000 for new construction, but even this modest sum was cut from the budget by Governor Brumbaugh's veto. Since the governor's action prevented the School of Engineering from enlarging its facilities until 1920 at the earliest, the board of trustees consented to erect one new unit for the School using the College's own limited resources. In October 1917, the trustees authorized the construction of Engineering Unit B, which was to house the Department of Industrial Engineering's wood and metal working shops. Work was begun the following year.

The trustees' approval of Unit B resulted in part from the extraordinary demands placed upon the School of Engineering by America's entry into World War I in April 1917. The United States was remarkably unprepared to assume a major role as an ally of Great Britain and France, in spite of the fact that the war had been raging in Europe since 1914. Hence, the armed services were ill-equipped for the task of raising a large, adequately trained fighting force in the short time President Woodrow Wilson had allotted. In the technical areas, especially, this training would have to be provided by civilian sources. Recognizing this necessity, President Sparks joined with college administrators across the nation in putting the facilities of their institutions at the disposal of the federal government as America mobilized for war.[4]

The war had an immediate impact on the School of Engineering. By June 1, 1917, the School had granted early dismissals from spring term to over 175 students, so that they might join the military directly or work on farms, in factories, on construction projects, or wherever else the war had caused labor shortages. The School also established several short, non-credit courses in order to better prepare those students who eventually would see service with the army or navy. Dean Sackett, an experienced yachtsman, taught a course in navigation, while Hugo Diemer and Professor of Mechanical Engineering Edward Bates conducted classes in ordnance and aviation, respectively. Over a hundred students enrolled in these free courses during April and May, undeterred by the extra workload these studies represented. With the conclusion of spring term, nearly all the engineering faculty took summer positions with railroads, utilities, construction companies, and in other industries where their technical knowledge would benefit the war effort. By June 30, ten faculty members, including department heads Diemer and Walker, had gone so far as to accept commissions in the armed forces.

Resumption of classes in September brought further changes to the School of Engineering. Most of the faculty who had joined the military were assigned more pressing duties away from the campus. Their absence, together with the School's inability to secure enough replacements in this time of national crisis, meant added responsibilities for that portion of the teaching staff that did remain. Dean Sackett, besides fulfilling his normal administrative obligations at the College, accepted a position on the War Department's Committee on Special Education and Training and in that capacity spent a good deal of his time in Washington. Eventually all of his colleagues in the School of Engineering joined him in defense-related activities of one kind or another. Initially the War Department preferred to have commissioned officers responsible for technical instruction on the nation's college campuses. At Penn State, John Orvis Keller, an instructor in the Department of Industrial Engineering, was made a second lieutenant and placed in charge of a course in "Storekeeping, Accounting, Continuous Inventory, Disbursing, and Transportation." Developed at the request of the Army Ordnance Department, the course was open to all undergraduate males. In April 1918 Penn State and hundreds of other colleges and universities began providing technical training for enlisted personnel in a continuous series of six- and eight-week courses. The bulk of this work fell to the School of Engineering and its regular civilian faculty. Harold Shattuck, for example, acting head of the Department of Civil Engineering, taught topographical surveying, while Charles Kinsloe offered a course in radiotelegraphy. After Keller was transferred to the ordnance school at Camp Hancock, Georgia,

Professor Bates continued some of his work in teaching men how to drive and repair cars and trucks. Other subjects taught by the engineering faculty included carpentry, blacksmithing, and sheet metal working, all offered on the recommendation or approval of the War Department. No objections were raised as the arrival of hundreds of soldiers in the spring and summer of 1918 began transforming Penn State from an institution of higher learning into essentially a trade school. No protests were voiced against the federal government's violation of traditional academic freedom and independence. Students and faculty in the School of Engineering, at least, believed that if making the College an instrument of the War Department would significantly increase the chances of military victory, then the institution's autonomy should be temporarily sacrificed.

To make available as many of the College's resources as possible for military instruction, President Sparks ended the spring semester on April 24, six weeks earlier than scheduled. This measure brought a little relief to the engineering faculty, who for the previous few weeks had been teaching classes for enlisted men in the mornings and evenings and civilian undergraduate classes during the afternoons. Until the resumption of the school year in September, the teaching staff devoted its attention exclusively to defense education. The war also brought radical changes to engineering extension and research. Those extension classes that were not directly beneficial to the war effort were cancelled. Most investigations at the Experiment Station were likewise halted. The only significant research that continued focused on ways to prevent dust explosions in grain elevators. Penn State was already the nation's leading center for research on this subject, thanks to the pioneer efforts of Benjamin Dedrick and the Pennsylvania State Millers Association. Thus the federal government turned to the School of Engineering for assistance in solving the explosion problem, which was especially serious in the Midwest, where elevators were filled to capacity to meet the overseas demand for food grains.

World War I placed a severe burden on the entire College, but nowhere was this strain felt more acutely than in the School of Engineering. By November 11, 1918, the date of the armistice, the School had given instruction to over 2500 troops and had in attendance on that date an additional 625 officers and men, as well as some 1100 undergraduates. (Legally speaking, even the undergraduates were members of the armed forces through compulsory enrollment in the newly formed Student Army Training Corps. The war ended before the SATC had any lasting impact on the School, however.) Engineering faculty and students were eager to return to the more relaxed conditions of peacetime. Even before the war's end, there were indications that a period of renewed growth lay ahead. Engineering Unit B was nearly ready for occupancy, with all but some interior furnishings in place. Sackett was busy drawing up plans for two more units and a transportation building—that elusive dream that his two predecessors had pursued in vain for so long. This time, the dream might finally become a reality, for the legislature showed signs that in recognition of the College's outstanding contributions to the war effort, it was prepared to be more generous with the upcoming biennial appropriations. As a gesture of its own confidence, the School of Engineering in the fall of 1918 admitted its largest freshman class ever.

Suddenly, this optimistic outlook vanished, as the School of Engineering found itself confronting the gravest crisis in its history. Just before seven o'clock on the balmy autumn night of November 25, 1918, passersby noticed flames licking from a second-story window of the woodworking shop, attached to the rear of the Main Engineering Building. The campus fire alarm, a banshee-like whistle at the power plant, sounded immediately, as students, soldiers, and townspeople came running to the scene. President Sparks, who had been entertaining dinner guests at the nearby President's House, stepped out onto

Inferno in the Main Engineering Building, as photographed from the lawn of Old Main. (Penn State Collection)

the veranda at the first wail of the alarm and could clearly make out the flames as they ate hungrily at the stacks of kiln-dried lumber stored in the shop. Sparks did not know it at the time, but several students were already attempting to beat back the flames. Entering the building by way of the front doors, they made their way to the second floor and ran down the hall in the direction of the wood shop. Upon reaching the corridor that connected the main building and the shop, the students grabbed a hose from a wall rack, coupled it to a hydrant in the corridor, and turned on the water. Not a drop emerged from the nozzle. Rotten with age, the hose had burst. The helpless students could do nothing now but retreat, filled with the sickening knowledge that the Main Engineering Building, for over twenty years the pride of the School of Engineering and indeed the crowning architectural gem of the entire College, was doomed.

Fire-fighting apparatus from the campus and from the borough of State College arrived on the scene minutes later, but by that time the fire had spread beyond control. The oil-soaked floor of the corridor linking the main building with the shop provided an ideal pathway for the flames. High winds threatened to carry the fire to even more structures. The volunteer firemen, reluctantly admitting their inability to save the main structure, turned their hoses on adjacent buildings instead. For a few hours, many spectators feared that even this struggle would be a hopeless one, as burning embers sailed skyward along College Avenue toward the eastern edge of the campus and the borough. Luckily, firemen from Tyrone and Bellefonte appeared on the scene around eight o'clock. Their efforts confined the blaze to the main building, where the inferno raged for another three hours, consuming the entire structure. Daylight found only a smoking hulk. Parts of the building's walls and chimneys remained upright, but the interior was completely gutted. The fire had engulfed the building so quickly that practically no furnishings or records were removed; loss was very nearly total. In addition to the Main Engineering Building and the woodshop, flames also ravaged the power plant, leaving the College without steam heat and electricity. The day after the conflagration, President Sparks made arrangements with the borough power company to supply electric current on a limited basis, but no heat could be obtained until makeshift repairs were performed on the damaged plant's boilers. All classes were suspended until December 5, when the heating equipment was finally returned to service.

Obviously, a resumption of classes did not mean a return to academic life as usual for many engineering students. The civil and mechanical engineering departments lost nearly everything in the fire,

Main Engineering Building shortly after the fire. (Penn State Collection)

since almost all of their laboratories, classrooms, and offices were located in the main building. The only materials of any real value retained by the Department of Civil Engineering were its surveying instruments, which were stored in Engineering Unit F. Ironically, the department for several years had been trying to obtain a different storage area, on the grounds that the all-wooden Unit F was a fire trap. The Department of Mechanical Engineering lost virtually everything. It did not even possess a locomotive, having sold the aging 4-4-0 to the Susquehanna River and Western Railroad in January 1918. The $1600 earned from the sale was deposited in a special fund to be used for the purchase of a more modern engine. Civil and mechanical engineering students would have had no place to go had not the recent departure of several hundred soldiers made vacant space available in the Mining Building and in Unit F. Offices and recitation areas were set up there pending a move to more permanent quarters. The problems arising from the absence of laboratories were not so easily resolved. The Department of Civil Engineering continued to utilize the College's sewerage treatment plant as a laboratory, but that was of little benefit to students who had no ambitions to become sanitary engineers. The Pennsylvania Railroad granted mechanical engineering students access to some of the test facilities at the road's Altoona works, but again, the assistance did not begin to satisfy all the department's needs. Bereft of any conventional laboratory apparatus on the campus, some mechanical engineering students received a good many barbs from their fellows for allegedly having to take practicums in "coal shoveling," "floor polishing," and "sweeping out the dynamometer car." The Department of Electrical Engineering was housed primarily in Units D and E and so escaped the fire unscathed. Likewise, students in architectural engineering had few adjustments to make, since their department was located in Unit F. The Department of Industrial Engineering lost its woodworking and lumber storage areas, but these were scheduled to be transferred to the new Unit B anyway. That building, the closest unit to the fire, had been upwind of the flames and suffered only some minor scorching.

College officials estimated the cost of the destruction caused by the fire to be about $300,000, of which $260,000 was ultimately recovered from insurance payments. Meeting soon after the fire, the board of trustees voted to use the insurance money to finish work on Unit B as soon as possible and to construct two more identical units, A and C. The trustees also authorized the addition of a third floor to Unit E. Construction proceeded swiftly, with Unit C ready for occupancy in the fall of 1919 and Unit A in January 1920. In Unit C was installed the Department of Industrial Engineering's metal working complex, including the forge and machine shops. The basement of

Undergraduates at work in the new Mechanical Engineering Laboratory. (Penn State Collection)

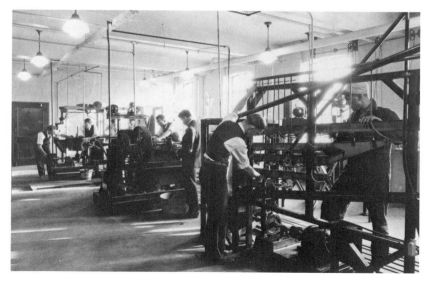

Unit A became the home of the materials testing laboratory, rebuilt from equipment salvaged from the main building, and what passed for the rudiments of a mechanical engineering laboratory. The upper floors contained the Department of Civil Engineering's masonry and highway laboratories and assorted offices and classrooms. This arrangement clearly shortchanged the Department of Mechanical Engineering. Department head Edward Fessenden and the rest of the mechanical engineering faculty argued strongly for a separate mechanical engineering building. Dean Sackett concurred. In fact, in his opinion even more buildings were needed. "It is important that a mechanical engineering laboratory and a Main Engineering Building be built at the earliest possible date, as there is every prospect of an increasing number of students in engineering," he told President Sparks a few months after the fire.[5]

The trustees granted the request for a new building for the mechanical engineering department and in November 1919 voted to use what remained of the insurance money to finance it. The new structure was to be erected along Burrowes Road, just north of the Bellefonte Central tracks and opposite the Beta Theta Pi fraternity house. It was to measure 60′ × 120′ and contain two stories and a partial basement. The College's architects, Day and Klauder of Philadelphia, prepared the design based on Professor Fessenden's study of the most modern college mechanical engineering laboratories east of the Mississippi River. Excavation of the foundation commenced late in 1919, and Penn State's own labor force handled most of the construction. Classes were first held in the new building during the spring of 1921. Original plans had called for a wing to be attached at each end of the central section to form an I-shaped structure; but since the central part alone cost $100,000, these additions had to await more prosperous times. The upper wing was to be a transportation laboratory, while the lower one was to contain offices and classrooms. Until the wings became reality, the central section would have to house just about everything the department possessed—which admittedly was not very much, in the year or two after the fire. Steam and internal combustion engines along with a couple of automobile chassis—all donated by private industry—occupied the first floor. The second floor, or gallery, was reserved for light machinery and bench work. Much discussion occurred regarding the possibility of rehabilitating the old Corliss steam engine that had been trapped in the fire, or of restoring it sufficiently to be displayed as a museum piece. Close inspection of the engine revealed that restoration of any sort would be an expensive proposition, and since the School of Engineering was already in dire need of money for other projects, the machine was eventually scrapped along with most of the other heavy equipment destroyed by the fire.

At a time when the College could barely muster enough money for an incomplete mechanical engineering laboratory, construction of a new main engineering building was out of the question until the legislature provided the funds, an event not likely to happen in the immediate future. The nation slipped into a brief but severe economic recession after the war, producing a fiscal climate in Harrisburg that was not conducive to making large expenditures for higher education. Perhaps if President Sparks could have spoken on behalf of the School of Engineering, the situation would not have become so critical. His warm, congenial personality had been one of Penn State's most valuable assets in its relations with Pennsylvania's lawmakers. If during his tenure the College had not received everything it requested of the Commonwealth, the legislators were nevertheless far more generous with their appropriations than they had been during the Atherton administration. Unfortunately, Sparks was unable to launch another public appeal on behalf of the institution. The crushing burden of war work that the President had shouldered was aggravated in October 1918 when his private secretary succumbed to nervous exhaustion, forcing him to assume responsibility for even the smallest details of his office. The burning of the Main Engineering Building in November and the administrative nightmares that followed the conflagration created a strain that no man could have been expected to bear, and in February 1919 President Sparks suffered a nervous breakdown. Time failed to bring a marked improvement in his health, so in January 1920 the trustees accepted his resignation. He spent the remaining four years of his life as a lecturer in American History at Penn State.

When the appropriation for 1921–23 was passed, it contained $2.4 million for Penn State, only two-thirds of the amount the College had requested. The School of Engineering received no money for new construction. State officials could hardly protest that they were unaware of the institution's dire need for more money. In May 1920, before the General Assembly had approved the upcoming biennial budget, a special committee of the Pennsylvania State Chamber of Commerce visited the Penn State campus to examine the College's facilities and discuss its requirements with school administrators. The report filed by this committee a short time later severely chastised the Commonwealth for rendering insufficient financial support over the last decade. It noted that since 1913, 3500 qualified students, 2500 of whom hoped to enter the School of Engineering, had been denied admission because Penn State lacked the resources to accommodate them. "A continuation of these conditions will rob the College of the public character derived from the foundations of the Morrill Act," said the committee, "and will arbitrarily restrict the invaluable service to the Commonwealth in the improvement of agriculture and the mechanic arts."[6]

The depressing state of affairs that the report went on to outline was in large measure responsible for the loss of more than a few distinguished faculty members. From the School of Engineering, easily the most prominent individual to depart was Hugo Diemer. Although officially head of the Department of Industrial Engineering until 1919, Diemer had spent the war years as a colonel in the service of the Army Ordnance Department. After the war, he accepted a position as Director of Management Courses for the LaSalle Extension University in Chicago. Until his death in 1939, Diemer continued his pioneering work in combining engineering and management studies and gave thousands of persons the opportunity to study industrial engineering while holding fulltime jobs. He was succeeded as industrial engineering department head by Edward J. Kunze. Another departmental headship had to be filled in 1919 when Roy Webber of the Department of Architectural Engineering resigned to become Penn State's Superintendent of Grounds and Buildings, a post he was to hold until his death in 1929. Succeeding Webber was Alfred Lawrence Kocher, who had been on the teaching staff of the department since 1912. A 1909 graduate of Stanford University, Kocher had received a master's degree from The Pennsylvania State College three years later.

By far the greatest number of changes in the engineering faculty occurred at the assistant professor and instructor levels. Not since the earliest years of the century had the turnover been so high as during these immediate post-war years. Nevertheless, the School of Engineering at this time received record numbers of requests for undergraduate admission. Enrollment climbed in response to this unprecedented demand, but the School could not expand its facilities to allow it to accept all qualified applicants. Over 200 qualified applicants had to be turned away from the 1919–20 academic year, for instance, while the following year over 400 were refused entry. The most distressing aspect of this situation was that virtually every one of these potential students was a resident of Pennsylvania. "It is a source of deep regret to the faculty and friends of the School of Engineering," Dean Sackett wrote in his annual report in 1921, "that it is not able to do the service required of it." Had the School been able to accept in 1919 and 1920 every applicant who had demonstrated a competency to master the challenges of an engineering education, it would have ranked third in

A 1920s view of the Bellefonte Central station, Engineering Units E, D, C, and B, and a portion of the power plant, from College Avenue. (Penn State Collection)

the nation in total enrollment. As matters stood, the School dropped from sixth place before the war to tenth by 1920.

The post-war years were not without a few positive accomplishments. The very fact that the School of Engineering could attract students in ever-increasing numbers represented proof that it still maintained a high standard of instruction, irrespective of the troubles that had beset it. The demands of employers likewise continued to outstrip the School's supply of graduates by a wide margin.

Engineering extension also flourished after the war, having been elevated from division to department status to put it on a more equal footing with the work being done in the other departments. One of extension's most significant achievements came in 1919 with the inauguration of a full-scale correspondence program. Prior to the war, a few correspondence courses had been offered, but they had languished for want of time and money on the part of the extension staff and the regular engineering faculty. Increased financial support and a small surplus of material resources left over from the war encouraged Extension Director Norman Miller to make another attempt to organize a correspondence section within his department. Among the first courses made available by mail were those dealing with engineering mathematics, shop engineering, mechanical drawing, steam engineering, gas engineering, and industrial management. At first these courses carried no credit and, according to the College catalog, were intended mainly for "machinists, foundrymen, pattern makers, carpenters, engine and boiler makers, power plant employees, and men in similar lines of employment" who desired to improve their skills at jobs they had already chosen. The School of Engineering nonetheless recognized that many correspondence students were formerly degree candidates who had withdrawn from school for one reason or another or were students who had never been able to attend college but who hoped eventually to earn a college degree. Consequently, beginning in 1920 the extension department began offering via correspondence over fifty courses that could be taken for credit. Such correspondence courses as Railway Mechanical Engineering 1 ("Locomotives"), Electrical Engineering 8 ("Dynamo Machinery"), and Architectural Engineering 19 ("History of Ancient Architecture"), were comparable in nearly every way to courses carrying the same names and numbers of credits that were offered through resident instruction. Correspondence students could choose from a list of subjects from every one of the engineering departments. By mid-1921, over 600 students were enrolled in these classes by mail, making Penn State the national leader in engineering correspondence instruction for college credit.

In cases where a large group of individuals (usually at least a dozen) wished to enroll in the same course, the extension department made every effort to provide an instructor who could teach the class in person rather than through correspondence. Another innovation involved a modification of the regular correspondence course and was called the "home study plan." Students were sent their lessons by mail and, after completing a specified number, met individually or as a group with a supervisor. The supervisor corrected their lessons, answered questions, offered advice, and generally supplied the "personal touch" that correspondence study lacked.

The growth in regular extension work matched that of the correspondence section. The extension department conducted classes for over 4300 students at thirty-two centers throughout Pennsylvania in 1920 and employed 127 teachers and supervisors. Pleased with the reception accorded the Allentown branch school, Professor Miller had founded a second one at Wilkes-Barre just before the war. It too thrived, and now Miller was making plans to set up even more of these permanent evening schools. The number of extension classes offered at all locations continued to expand and become more sophisticated in order to meet the changing needs of the students.

Meanwhile, work at the Engineering Experiment Station resumed and soon surpassed its pre-war level. Just as before the war, much of the activity centered around problems of heat transmission and insulation, with A.J. Wood continuing in general charge of the station. The thermal laboratories, although ten years old, still constituted one of the best facilities in the country for research in heat transfer in building materials. Work done there had already proven the value of air spaces as insulators, with Professor Wood himself utilizing recent discoveries to develop a more efficient means of insulating railway refrigerator cars. In recognition of the station's achievements, the American Society of Heating and Ventilating Engineers awarded it $500 in 1920 to underwrite further research in the area of insulating techniques. This grant pointed up the major difficulty that handicapped work at the Experiment Station: insufficient funds to support research in a broad range of categories. The station had to turn down countless requests from companies and individuals to pursue research in many areas, since only a tiny fraction of these requests were accompanied by supporting monies. The small staff, all of whom were affiliated with the station on a part-time basis, simply did not have the resources to do independent work. The many appeals received for information concerning subjects already under investigation at the station or about which faculty members had some knowledge were more easily handled. In 1921 Dean Sackett established the Engineering Service Bureau, consisting of himself and all department heads. The bureau answered hundreds of queries by mail, free of charge, on a broad variety of topics, ranging from how to select proper acoustics for a church to how to build a concrete pig house.

The most noteworthy development in student affairs during the early 1920s was the appearance of the student body's second attempt to issue a self-sustaining engineering periodical. Titled the *Penn State Engineer,* the first issue came off the press in May 1920. Paul B. Kapp ('20) served as the first editor and Elton Walker as faculty advisor. Vowing "to create and awaken new interest between the students of the Engineering School and the alumni of the School," the *Penn State Engineer* was the product of a newly formed Combined Engineering Society, a student organization. None of the individual student engineering societies had the capacity, financial or otherwise, to support a major publishing venture, so a cooperative effort was imperative if Penn State were to join the growing ranks of schools having their own student engineering journals. Yet convincing all of the various departmental societies to work together was no easy task, as the absence of cooperation that had preceded the demise of *The Engineer* in 1910 had proven. This time, because the individual societies were more willing to work together in the Combined Society, the friction and disunity of an earlier day did not reappear, and the *Penn State Engineer* survived. Issued as a quarterly beginning in the fall of 1921, it carried the usual array of informative articles by students, faculty, distinguished alumni, and prominent engineers. Unlike *The Engineer,* the new publication from its very first number offered more detailed coverage of student activities and achievements and featured far more advertisements from business and engineering firms. The ads helped keep the annual subscription price down to an affordable one dollar per year.

An Attempt for Renewed Growth

An end to the stagnation that had set in after the fire and the illness of President Sparks seemed near as the board of trustees completed the process of selecting the College's ninth president. The new chief executive was John Martin Thomas, for the previous twelve years president of Middlebury College, a small Vermont liberal arts school. An 1890 graduate of Middlebury, Thomas had spent fourteen years as a Presbyterian minister before returning to his alma mater as its head. His lack of experience with technical education did not blind him to the crucial role engineering had played in the evolution of Penn State. "The strength of the Engineering School helps every other department of the institution," he proclaimed in his inaugural address. "We have more students today in agriculture and mining than we would have if the College were devoted entirely to agriculture or mining." Clearly, the new president had no intention of slighting engineering in his plans to bring a period of new growth to The Pennsylvania State College. Thomas began his new duties at the College with enthusiasm. Noting that "you cannot do in Pennsylvania with one dollar what it would take five dollars to do in any other state," Thomas, along with the trustees and influential alumni, began mapping strategy for a giant fund-raising campaign to begin in the fall of 1922. The goal was eventually set at two million dollars, most of which was to be spent on new buildings.[7]

No one doubted that the School of Engineering required more classrooms, laboratories, and offices. Speaking at an industrial relations conference held on the campus in October 1921, President Thomas observed with more than a hint of bitterness that three years had passed since the burning of the Main Engineering Building and still no immediate prospect existed for erecting a replacement structure. This inaction, in his estimation, could be traced to inadequate state support. Thomas accused the legislature and Governor William Sproul of "al-

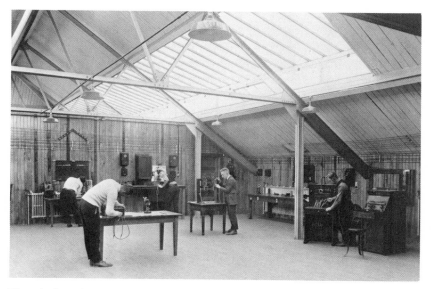

The telephone laboratory on the third floor of Unit D, about 1924. (Penn State Collection)

most criminal shortsightedness and economic waste," in view of the high degree of industrialization prevalent in the Commonwealth and the proven contribution of Penn State engineers. (Over half of the School of Engineering's graduates lived and worked in Pennsylvania at that time.) These circumstances, Thomas said, made all the more necessary a vigorous campaign to convince alumni and friends of the College to contribute generously to a building fund.[8] While no definite plans were formulated as to how much of this fund would be channeled to the School of Engineering or for what specific purposes, the President and Dean Sackett agreed that a new main engineering building should receive a high priority. The Departments of Civil Engineering and Mechanical Engineering were still scattered throughout three separate buildings, even though the mechanical engineers did enjoy the luxury of having a spacious new laboratory. The Department of Architectural Engineering had also expanded to such an extent that Unit F alone was no longer sufficient to accommodate it. All three of these departments would be likely candidates to occupy a new main building. A new electrical engineering building would rate second priority. The electrical laboratories in Units D and E were overcrowded and did not give the 600 or so students who were using these facilities enough space to work efficiently. The electrochemical laboratory in

Unit E produced fumes that were irritating and often dangerous to occupants of the entire building. Both units, with their wooden walls and floors, presented a fire hazard that could not be ignored. Final plans for these and additional buildings would hinge on just how much cash could be generated by the forthcoming Emergency Building Fund Campaign, as the drive was termed.

More buildings would not solve all the School of Engineering's most pressing problems, of course. Additional improvements were badly needed in other areas as well. That point was made abundantly plain in May 1922, as another two-day industrial conference gathered at the College. Composed of representatives from more than forty companies that maintained headquarters or significant portions of their operations in the Commonwealth, the conference met "for the purpose of investigating the instruction, equipment, buildings, and finances of the School of Engineering." If President Thomas and Dean Sackett, prime movers of the conference, expected the investigation to highlight the School's demand for more and better teaching resources, they were not to be disappointed.

In their final report, the conference attendants concluded that the School's "instruction is sound, and gives due weight to character, citizenship, and service to the State." Their assessment of the School's laboratory apparatus and other instructional equipment was not so favorable. The investigating team found it to be

> fair in quality, but insufficient to handle the present student body effectively. The equipment is totally inadequate to provide for the instruction of all the worthy and well-qualified young people of Pennsylvania who are applying for entrance to this, the State institution. We urge that immediate additions to equipment be made on a scale sufficient to provide for the desirable growth of the School, as well as to bring the present level of instruction up to proper levels of efficiency.[9]

The report "heartily endorsed" plans for a new main engineering building and a new electrical engineering building as "an absolute necessity if the institution is to perform the service which the State and federal governments have designated it to do." The matter of salaries drew the sharpest criticism. "We are shocked at the meager appropriation and low salaries paid by one of the largest and most important schools of engineering in the United States," declared the conferees. "The teacher is the most important factor in technical education. . . . The state owes an obligation to its teachers that it has not paid." The report did not confine itself to an investigation of resident instruction, but delved into the operations of the Engineering Experiment Station and engineering extension, too. Work in both areas was deemed to be satisfactory. Extension, the conference report affirmed, "has helped thousands of men and women to become better workmen, has been an important factor in their promotion, and has helped them to be better citizens." Engineering research was an equally vital function of a state institution. Pennsylvania's small businesses and industries relied heavily on the Experiment Station, for the limited resources of these firms did not allow them to operate their own laboratories in which they could investigate some new process or invention. The activities of Professor Wood and his associates at the station helped the Commonwealth's citizens and industries to develop new sources of wealth and to raise the standard of living. These were legitimate functions of a state institution, according to the report; research deserved more public encouragement than it had thus far received.

The representatives of the industrial conference concluded their report by urging the Commonwealth to allocate $1.8 million to the School of Engineering during the upcoming biennial appropriation period (1923–25), if it wished to correct the deficiencies highlighted by the conference and make the School the finest possible institution for engineering education. About $1.3 million of this amount should be allocated to construct and furnish new buildings and modernize old ones. A hundred thousand dollars should be spent on upgrading the facilities of the Experiment Station, with the remaining money being used to increase the salaries of the teaching staff and to hire more instructors.

Dean Sackett vigorously seconded these suggestions. In his annual report of 1921–22, he pointed out that seven of the eight faculty members who had resigned during the year had given inadequate compensation as the primary reason for their leaving. One of these seven was Edward Fessenden, head of the Department of Mechanical Engineering. He left Penn State to accept a position of lesser responsibility at Rensselaer Polytechnic Institute, which had offered to double the salary he was then earning. His resignation "proved more emphatically than ever," Sackett insisted, "that in engineering we must pay increased salaries to good teachers." To fill the vacancy created by the departure of Fessenden, A. J. Wood was named as the new head of the Department of Mechanical Engineering, while Associate Professor of Mechanical Engineering Fred G. Hechler succeeded Wood as supervisor of the Experiment Station. Hechler, a 1908 mechanical engineering graduate of the University of Missouri, had taught at Rensselaer Polytechnic Institute for eight years before joining the engineering research staff of the Naval Experiment Station at Annapolis in 1916. The only vacancy on the faculty that did not stem at least in part from low pay was that caused by Professor Kunze, who retired as head of the Department of Industrial Engineering. J.O. Keller, who had first won recognition for his outstanding work with the Army Ordnance School during World War I, was appointed to succeed him. After his military service, Keller had taught in the mechanical engineering department of Iowa State College for three years. He returned to Penn State in 1921 as Associate Professor of Industrial Engineering to assist in establishing that department's model factory, which manufactured and sold wood products, such as cedar chests, bureaus, and porch swings, and produced much of the furniture used in the College's offices and classrooms. The factory gave industrial engineering students the opportunity to gain practical experience in production methods, conducting time and motion studies, cost accounting, and marketing.

Until the Emergency Building Fund Campaign reached fruition or until the governor and legislature could be convinced to lend increased financial support to the College, very little could be done to alleviate most of the problems plaguing the School of Engineering. Apparently the report of the industrial conference and the inordinately large number of resignations in 1921–22 did have some positive effects within the College administration, however. By carefully shifting and transferring funds that otherwise would have gone to pay for maintenance of physical plant and the acquisition of new equipment, President Thomas in 1922 was able to secure a modest pay hike for the faculty of the School of Engineering. The largest increases went to those persons holding senior appointments. The average annual salary of a department head rose from $2800 to $3500, for instance, and that of a professor from $2300 to $2950, while an instructor's average pay was boosted barely $400, rising from $1354 to $1756. These increases put Penn State on a par with most other land-grant institutions in the realm of engineering salaries.[10] Unfortunately, the decrease in funds for new equipment and for plant maintenance would inevitably take its toll and would have to be made up in the not too distant future.

The College could expect little help from Harrisburg. The General Assembly approved a $2.9 million appropriation for 1923–25, but newly elected Governor Gifford Pinchot, preaching a gospel of financial restraint, reduced this sum by veto to $2.1 million, $300,000 less than the appropriation for 1921–23. The legislature chose not to try to override the veto, leaving President Thomas with no recourse but to rely on Penn State's own fund drive to obtain some relief from the overcrowded conditions that prevailed campuswide. The Emergency Building Fund Campaign did not attain the success that College officials had hoped for. After expenses, Penn State netted only $1.2 million by June 1924, when the campaign ended, and even much of this sum was in the form of pledges, which could not be redeemed immediately. A sizeable portion of the cash already on hand was used to build an infirmary and men's and women's dormitories. The School of Engineering received not a dollar of these building funds. Whether Dean Sackett expected to share in the proceeds of the campaign, toward which many engineering alumni had made substantial contributions, is not known. If he did, he gave no outward sign that he was dissatisfied or that he believed his School had been unfairly treated. For this attitude he won praise from President Thomas. "Dean Sackett and his associates have not spent their time in bewailing the limitations forced upon them," Thomas said in October 1924, "but in making utmost use of the facilities at hand."[11]

The financial problems vexing the School of Engineering during the early 1920s should not obscure other more constructive events occurring at that time. Important curricular changes were made, for example, in two of the engineering departments. In 1920 the Department of Industrial and Fine Arts, formerly under the jurisdiction of the School of Liberal Arts, was transferred to the School of Engineering. This department was established in 1890 as the Department of Industrial Design to teach, according to the College catalog of that year, "free-hand, industrial, and mechanical drawing" to students in both technical and non-technical courses. Graphics then comprised a sizeable portion of the engineering curriculum, particularly for juniors and seniors, whom the catalog stated were to receive instruction in "mechanical and engineering drawing, copying from the flat and from models, topographical sketches and maps, isometric and plane projections, plotting surveys and triangulations, lettering, map drawing, coloring and shadowing." Under its first head, Anna E. Redifer, the department was quick to branch out into the fine arts, also. By the eve of World War I, it was offering work in painting, interior decoration, pottery, costume design, and art history and appreciation. Merging industrial and fine arts with architectural engineering enhanced the School of Engineering's control of graphics instruction but created the incongruity of having technologists exercise supervision over education in the arts. In 1922 the trustees agreed to the division of the architectural engineering curriculum into two distinct baccalaureate programs, architectural engineering and architecture, both within a newly renamed Department of Architecture. The new course in architecture combined the essentials of a liberal education with strictly professional work and stressed the creative planning and design of structures, along with architectural history. The ultimate goal of the course was to prepare the student for independent practice as an architect. The architectural engineering course continued to emphasize the technical aspects of structural engineering, requiring much course work in mathematics and engineering mechanics. It aimed to train students to supervise the construction of buildings and other structures. In 1923 the board of trustees approved the creation of a School of Education. This new entity drew most of its components from the other schools. The course in industrial education, heretofore a part of the Department of Industrial Engineering, became the Department of Industrial Education within the School of Education. William P. Loomis ('10, electrical engineering) was named to head the department.

Equally notable as the administrative changes that did take place was one that did not. In 1922, when Elwood S. Moore resigned as Dean of the School of Mines, Dean Sackett made a strong appeal to President Thomas to dissolve that school and transfer most of its academic responsibilities to the School of Engineering. Sackett claimed that, largely through no fault of its own, the mining school had always suffered from inadequate financial support. The lack of sufficient space, equipment, and qualified instructors inevitably caused the School of Mines to have the smallest enrollment of any of the College's schools. Incorporation into the School of Engineering could change that dreary picture, Sackett argued, by enabling the various engineering departments to present a more united front in their attempts to win greater funding. In addition, the Dean of Engineering cited the frequently poor performance shown by many mining students when taking courses in the School of Engineering to show that the academic standards of the School of Mines could be measurably improved. Objections from faculty and students within the mining school, together with preoccupation with problems elsewhere, caused President Thomas to look with disfavor on Sackett's proposal, however, and the matter was quietly dropped.[12]

Generating far more popular interest than the doings of the academic hierarchy were the activities of the College's new radio station. Students in the Department of Electrical Engineering tried unsuccessfully at the end of World War I to convert the old wireless station to radio phone operation. Then in 1921, Professor of Electric Railway Engineering Eugene C. Woodruff, who also had a fascination for the emerging field of electronic communications, designed a new transmitter for the station. Several Pittsburgh area alumni subsequently donated a thousand dollars' worth of new equipment, which was assembled according to Woodruff's plans by Gilbert H. Crossley, an instructor in electrical engineering. The station began broadcasting in January 1922, having been assigned call letters WPAB and a license to broadcast at 500 watts, powerful enough for its signals to reach all parts of the Commonwealth. Penn State's Department of Public Information did most of what little formal programming existed, since the station was on the air only a few hours each week. Talks by faculty and other College representatives were featured. Technical and financial difficulties plagued operations throughout the spring semester, so after the station signed off just before commencement in June 1922, it remained silent for nearly two more years while students and faculty tried to remedy its troubles. The Department of Electrical Engineering continued to operate its experimental station, 8XE, but it had been converted to short wave, and the College was unable to utilize it as an official voice for public affairs purposes. In January 1924, WPAB—or more accurately WPSC, as the Department of Commerce agreed to designate the station—returned to the air, but once again was forced to cease broadcasting at the end of the semester. It operated only sporadically until early in 1927, when its transmitting gear was completely overhauled or replaced. From then on, WPSC broadcast regularly, presenting live coverage of athletic events, Sunday chapel services, and musical concerts, as well as the usual selection of faculty lectures. Donald M. Cresswell, head of the Department of Public Information, served as program director, and Gilbert Crossley acted as station manager. Both Crossley and Woodruff taught undergraduate courses in radio engineering.

The 1920s also witnessed a considerable increase in the activities of student organizations. Student sections of such professional organizations as the American Society of Civil Engineers and the American Institute of Electrical Engineers replaced the independent departmental societies. Honorary societies for engineering students made their first appearances on the Penn State campus, too. Electrical engineering students, for example, were eligible for membership in Eta Kappa Nu, while architectural engineering students could join Scarab. Sigma Tau and Tau Beta Pi welcomed all engineering students who met the stringent academic and personal admission standards. Freshman and sophomore civil, mechanical, and industrial engineering students who participated in the College's Reserve Officer Training Corps formed an independent engineering battalion within that organization. The Combined Engineering Society's offspring, the *Penn State Engineer,* regularly won recognition as one of the better student engineering periodicals in the country and enjoyed widespread popularity among students and alumni. Its appeal undoubtedly came in spite of, rather than because of, the periodic homilies that the editors offered for the moral and academic improvement of the student body. "One of the faults of the average engineering student is that too much of his energy is directed into channels which do not enable him to produce the best and most useful results," observed a student editor in March 1923. "Excessive time spent in movies, pool, and aimless discussions that lead nowhere and destroy one's mental ability rather than increase it, is an example of grossly misapplied energy. Like the stray power of an electric motor, it is wasted and can never be recovered."

The *Penn State Engineer* also devoted much space to recounting highlights of the annual inspection trips taken by junior and senior engineering students, usually during the Easter recess of the spring semester. These outings, an annual feature of the Engineering School since the turn of the century, lasted four or five days and were supposed to acquaint the students with the practical applications of the technical information they had absorbed in classroom and laboratory. The specific itinerary of each year's trip varied according to department and student interest, but power plants, railroad installations, steel mills, water works, and factories were regulars on the agenda. Civil engineers were also likely to visit a few construction projects, and architectural students preferred to tour famous skyscrapers and historic buildings. Regardless of a student's major, these inspection trips were immensely popular—and not necessarily because of their "educational" value, if reports of the trips published in the *Penn State Engineer* are to be believed. "Inspecting, of course, was the last thing thought of on such an occasion," a civil engineering student confessed in summarizing his department's annual inspection tour. "The thought of a chance to get out and away from the tiresome work of the classroom and to see the big cities, to see something new, something besides books and profs, and to hear anything but the Old Main bell; these and many other things caused extensive plans to be laid, and that old line to Dad about being broke to be revised again."[13]

Civil engineering students received an additional opportunity to apply their classroom knowledge with the advent of an eight-week summer camp. Prior to 1920, undergraduates in the various civil engineering courses were required to take a two-week surveying practicum, held entirely on the College campus, at the end of their junior year. Because of the large numbers of students enrolled in these practicums, however, and because the campus itself offered a limited number of experiences, the Department of Civil Engineering began searching for a more satisfactory site. Summer practicums were conducted on a trial basis at nearby sites at Lamar and in Penn's Valley. In 1923, the department found a more suitable location at Kellyburg in Lycoming County, about fifteen miles north of Williamsport. Each June, beginning in that year, about 50 civil engineering students entrained at Lemont for the three-hour ride to the camp, where they lived in tents and otherwise communed with nature for the next two months. Members of the departmental faculty supervised the instruction and generally acted as chaperones. For several weeks during the encampment, the engineers shared quarters with students from Penn State's Department of Forestry, who were also taking their annual practicums. Despite the military regimen to which the students were subjected during camp, they took pleasure in the opportunity to sharpen their skills under practical conditions. As in the case of the junior and senior inspection trips, these outings received detailed attention in the *Penn State Engineer* and many fond remembrances in *La Vie*.

No slackening of employers' demands for the College's engineering graduates occurred during the 1920s. Students in some of the engineering disciplines commonly received five or six job offers months before commencement. Every department routinely received many more requests for job applicants than it had graduating students. The demands for its alumni was only one index of the School of Engineering's success. Another was the extraordinarily large segment of graduates who remained in the engineering profession for all or most of their careers. Up to and including the Class of 1923, the School had graduated over 2400 students since the establishment of a four-year course in civil engineering in 1881. A survey of these alumni taken during the winter of 1923–24 revealed that, of the approximately 700 respondents, over 95 percent were employed in engineering positions, with about half holding managerial or supervisory posts. This figure compared very favorably with the national average, which was estimated to be around 65 percent. Also comparing well with the nationwide norm was the average starting salary ($1600) for a Penn State engineer. The survey showed that alumni with five years of engineering experience earned an average annual salary of $3110, with eight years, $4710, and with 17 years, $6840. (These figures make an interesting contrast with the salaries received by the College's engineering faculty, as noted previously.) About 60 percent of those responding resided in Pennsylvania, a conclusive demonstration of the value of engineering education to the economic well-being of the Commonwealth.[14]

Indeed, throughout the 1920s Penn State regularly conferred about a third of all baccalaureate degrees in engineering given by

Pennsylvania's colleges and universities. In 1923, a typical year, The Pennsylvania State College awarded 258 such degrees; the Carnegie Institute of Technology, 120; Lehigh University, 108; the University of Pennsylvania, 107; the University of Pittsburgh, 91; Lafayette College, 39; and Bucknell University, 32.[15] Penn State's School of Engineering was the only institution in the Commonwealth to have baccalaureate curriculums in railway mechanical engineering, electrochemical engineering, milling engineering, and architectural engineering. Its mechanical and electrical engineering departments were by far the largest in Pennsylvania in total undergraduate enrollment, with the Department of Electrical Engineering being the largest in the nation. The importance of a large force of well-trained engineers to a state that produced one-half (by dollar value) of the nation's industrial output and led the nation in the manufacture and use of steel, electrical, and transportation products was obvious.

Activities of individual faculty members also brought distinction to the School of Engineering. In 1924, Arthur J. Wood revised his book on locomotive operation, and it soon supplanted the earlier edition as the authoritative text for railway mechanical engineering courses in many American engineering schools. Benjamin Dedrick authored an equally distinguished work, *Practical Milling,* which immediately won acclaim among educators and industrial leaders as the standard textbook for milling engineering students and as a valuable reference volume for mill operators. Alfred Kocher, head of the Department of Architecture, won national fame for his investigations of Pennsylvania's colonial architecture. In a project spanning several years, Kocher and several of his students compiled an illustrated catalog of the Commonwealth's most important structures dating from the colonial era. After carefully examining each building's historical value and evaluating its current structural condition, Kocher selected those most worthy of saving for posterity and suggested ways by which they could be restored to their original character. His pioneer effort in the field of historic preservation sparked national interest in the subject. Robert Sackett, too, found himself in the national spotlight. Throughout his tenure as dean, he took an active part in advancing the case for professionally trained engineers and the institutions that produced them. After serving on various committees of the Society for the Promotion of Engineering Education in the early 1920s, he was elected president of that organization in 1927. In that capacity he spoke forcefully not only on behalf of The Pennsylvania State College, but for engineering schools across the country.

Frederick George Hechler, director of the Experiment Station, was indirectly responsible for introducing a second major field of engineering research at the College. Soon after coming to Penn State, Hechler caught A. J. Wood's enthusiasm for research in heat transfer and in 1924 helped develop the world's first reliable heat meter. This device measured even the most minute flow of heat through building materials. It was particularly useful to the cold storage and building construction industries, which responded to the progress made at the station by making additional monetary contributions to encourage further innovations. Meanwhile, Hechler desired the station to branch out into other research fields and in 1923 invited Dr. Paul Schweitzer to join his staff. Schweitzer, a mechanical engineering graduate of the Royal Hungarian Technical Institute and a former artillery officer in the Austro-Hungarian army, had only recently arrived in the United States. He was brought to Hechler's attention by A. J. Wood, who had been impressed by an article Schweitzer had published in a technical journal. "A page of topics was put in front of me," Schweitzer recalled of his first meeting with Hechler. "What would I like to do research on? I remember I designated pneumatic transportation—mail and packages and so forth—and I added as a whim that more than anything else I would like to work on diesel engines," a topic not on the list. Schweitzer's response intrigued Hechler, who had conducted some inquiries of his own in diesel engineering while at the Naval Experiment Station. Before he could assent to Schweitzer's request, however, he had to obtain the consent of Dean Sackett, since diesel engineering was not among the fields being considered for investigation at the station. Sackett, aware that no American college or university had undertaken systematic research on diesel engines, saw the opportunity for Penn State to pioneer in the field. He told Hechler to have Schweitzer "go full steam ahead" on the diesel project.[16] From this curiously inappropriate metaphor originated the Engineering Experiment Station's diesel laboratory, at first hardly more than a couple of diesel engines in the Mechanical Engineering Laboratory, but soon to become an internationally acclaimed center for diesel research.

The work of the Department of Engineering Extension was no less important than that of the Experiment Station. By 1921, eleven years after its founding, the extension department had given instruction to over 10,000 residents of Pennsylvania at a total cost to the College of about $75,000. Salaries of the principal extension faculty accounted for most of this sum, since in other respects extension activities were expected to be self-supporting. With an enrollment of some 7500 students in 1921, Penn State ranked fourth in the nation in engineering extension. Only the University of Wisconsin (where Louis Reber was directing extension activities), the state of Massachusetts (where James A. Moyer was in charge), and Iowa State College ranked higher. As part of his proposal to streamline the administrative structure of the College, President Thomas weighed the possibility of consolidating the extension work of the Schools of Engineering, Mines, and Liberal Arts into a single division. (Agriculture was to remain separate.) Norman Miller, department head of engineering extension, cautioned against such an action. "Any change involving the identity of the department would undoubtedly involve a loss of the hearty cooperation of the engineering faculty," he stated in a letter to Thomas in November 1921. The faculty had always supported extension work "to the limit," Miller asserted, "largely because it was *engineering* extension." He warned that achieving the same degree of cooperation would be "difficult if not impossible" should extension be removed from the Engineering School. In addition, Miller pointed out that engaging in extension duties on a part-time basis kept the faculty in close contact with practical applications of their subjects. This contact in turn helped the faculty keep practical matters uppermost in the classroom, "a thing which is often neglected in engineering colleges," according to Miller.[17] Spokesmen for the extension divisions of the other schools voiced similar arguments. President Thomas, unwilling to embroil himself in a bureaucratic struggle at such a crucial time in the institution's history, opted to remain with the status quo.

Its independence assured, engineering extension continued to prosper under Miller's direction in the early 1920s. In 1923, the department established a new school at Scranton, the third (after Wilkes-Barre and Allentown) that it owned and operated on a permanent basis. Engineering extension classes also began that same year at the Rockview State Penitentiary, five miles from the College, where 130 inmates were enrolled. The extension department assumed all costs of the prison program, but in every other case, most of the expenses of conducting the classes were borne either by the students or by the sponsoring organizations, most of which by this time were industrial firms. By 1924, the number of students that had taken engineering extension work climbed to over 60,000. A year later Miller resigned to become director of extension for Rutgers University. Named to succeed him was J. O. Keller. Charles W. Beese, a graduate of Iowa State and a member of the Penn State engineering faculty since 1922, replaced Keller as the head of the Department of Industrial Engineering. The 33-year-old Beese, a specialist in mass production methods, was the School of Engineering's youngest department head and gave much promise for future distinction.

Reaffirmation of the Importance of Engineering

In a report issued in 1924, the United States Bureau of Education estimated that the nation would need 500,000 engineers of all types by the end of the decade. If current rates of increase in enrollment in America's engineering schools continued, these institutions could be expected to have trained no more than 125,000 engineers by 1930. And if even as many as 250,000 engineers—a most optimistic figure—came up through the ranks rather than from colleges, 125,000 additional engineers would still be required. Obviously, the country's engineering schools had to find ways and means of vastly increasing the number of their graduates, without diluting the quality of professional education each student received. If the supply of engineers could not keep pace with the demand, Pennsylvania, with its heavy dependence on technology and industry, would be one of those states most adversely affected. "By far the greater number of men turned away by The Pennsylvania State College have been applicants for courses in engineering, the sons of Pennsylvanians, potential engineers for the future," Dean Sackett warned in 1924. "These men are being lost to industry. They are being diverted to other callings and to other states."[18] Yet without significant increases in financial assistance, Sackett was powerless to avert the misfortune that eventually might befall the Commonwealth.

As far as John Thomas was concerned, the uncertain outlook for engineering matched that for the College as a whole. As one of the major objectives of his administration Thomas had proposed that Penn State become a state university, owned and operated by the Commonwealth in the fashion of the great state universities of the Midwest. With Penn State thus the capstone of Pennsylvania's free public school system, Thomas argued, the legislature and governor had an obligation to see that its financial needs were met before appropriating any money to other institutions of higher learning. (The University of Pittsburgh, the University of Pennsylvania, and Temple University were then regular recipients of state funds.) He further suggested that Pennsylvania's constitution be amended to allow the Commonwealth to issue $8 million in long-term bonds on Penn State's behalf. An amendment was necessary in order to raise the constitutional limit on Pennsylvania's indebtedness and would have to be ratified by the state's electorate before it could be implemented. Embittered by the criticism his plans had provoked in the legislature and among educators at other schools and discouraged by the seemingly endless amount of time required to secure the passage of the bond issue, the President submitted his resignation in June 1925. In the letter announcing his decision to the trustees, he wrote that "for many years legislative appropriations to The Pennsylvania State College have been far below the careful estimates prepared by the institution," and there appeared to be little likelihood of any change. Thomas was leaving Penn State to become president of Rutgers University, where, he said, he expected to be of more "constructive service."

Just before Thomas's departure, Dean Sackett submitted to him a long-range building plan for the School of Engineering. "It is safe to say," the Dean commented, "that no institution of repute and with the engineering enrollment of The Pennsylvania State College has as small investment in buildings."[19] To correct this situation, the School urgently needed a new main building, the plans for which had already been prepared by the Department of Architecture. The new structure, to be three stories high and some 275 feet in length, was to occupy the same site as its predecessor. (A heap of bricks surrounded by a board fence still remained as a depressing reminder of the 1918 fire.) The new main building would contain classrooms and offices for the Departments of Civil Engineering and Architectural Engineering, as well as a hydraulic laboratory, a facility the civil engineers had been without since the fire. A lecture hall large enough to seat 250 people was to be located at one end of the building. Plans for the other engineering buildings were not as detailed, but construction priorities and costs had been arranged according to the following schedule: Main Engineering Building, $600,000; Electrical Engineering Building, $250,000; Engineering Experiment Station, $175,000; Transportation Building, $150,000; completion of two wings for the Mechanical Engineering Laboratory, $125,000; and Foundry, $100,000.

The relatively high priority awarded a separate building to house the Engineering Experiment Station represented a noteworthy change from previous building plans submitted by the School of Engineering. The increased emphasis on the station stemmed partly from the ever greater amounts of research it was called upon to undertake for various industries and governmental bodies and from the need to provide adequate research facilities for graduate students. The number of students studying for advanced degrees in engineering at Penn State was low in comparison to the numbers enrolled at other first-rate engineering institutions. Dean Sackett attributed this imbalance primarily to the absence of comprehensive facilities which graduate students could utilize for research purposes, although lack of faculty interest in conducting and supervising research was another inhibiting factor. The demands of undergraduate instruction severely limited the amount of time the teaching staff could devote to supervising graduate work—the teaching load averaged 18 class hours per week—but regardless of faculty availability, few graduate students would ever be attracted to an institution having inferior research facilities.

Additional buildings and the equipment with which to furnish them would not cure all the ills of the School of Engineering. "Salaries are still far below those of competing institutions," Sackett noted in his annual report for 1926–27. Low pay still caused an abnormally high turnover among the lower faculty ranks, although the situation had improved appreciably over the past several years. The most serious loss from the teaching staff came when Professor Kocher left to become dean of the McIntire School of Art and Architecture at the University of Virginia. He also accepted an appointment to the advisory committee that guided the restoration of Colonial Williamsburg, a project just then getting started, and won much praise for his work in that capacity.

It would remain for John Martin Thomas's successor to consider Dean Sackett's building proposals. Not until September 1926 did the trustees select that individual. He was Ralph Dorn Hetzel, for many years head of the political science department at Oregon State College and since 1917 president of the University of New Hampshire. The 43-year-old Hetzel had no illusions concerning the gravity of the problems he faced at Penn State. After having spent only a few months at his new job, he readily admitted that he presided over a physical plant "which is inadequate in size, and in part exhausted, and in part unfit and dangerous for occupancy."[20] Fortunately, in contrast to his predecessor, Hetzel was finally able to do something to remedy this problem. Favored on the one hand by a surplus in the state treasury and on the other by a governor, John S. Fisher, who was more sympathetic to the needs of higher education than any of the Commonwealth's chief executives since John K. Tener, the College received a building appropriation for 1927–29 of one million dollars.

The board of trustees voted to use $300,000 of this amount to finance construction of a new main engineering building. The new structure was to be built along the lines suggested in Sackett's plan of 1925, with a few modifications recommended by Day and Klauder, the College architects. Ground was broken on the site of the old Main Engineering Building in August 1928, with Berkebile Brothers of Johnstown acting as the general contractor. Composed of red brick trimmed with Indiana limestone, the building was designed to harmonize with the Georgian architecture of the other campus buildings recently completed or still under construction (Recreation Hall, a service building, an infirmary, and portions of today's West Halls dormitories). Measur-

Architect's rendering of the new Main Engineering Building. (Penn State Collection)

ing 60' × 230' overall, the new Main Engineering Building contained three floors and a basement. The long-awaited hydraulic laboratory occupied the basement; classrooms and offices of the Department of Civil Engineering were located on the first two floors. The third floor provided quarters for the Department of Architecture's drafting rooms and offices. Plans called for the addition at a later date of north and south wings to the building housing a lecture hall, a library, club rooms, and more classrooms. The absence of a main engineering building probably accounted more than any other factor for the more or less stagnant enrollment of the School throughout the 1920s, a stagnation that occurred at the very time that the total number of undergraduate students attending the College increased substantially. Total College enrollment rose from 3150 in 1920–21 to 3780 in 1927–28, while the School of Engineering's undergraduate population grew from 1125 to 1194 during that same time. The School of Liberal Arts and the newly created (1923) School of Education absorbed a sizeable portion of the growth of the student body.

The decision to construct the long-awaited Main Engineering Building had an immediate and positive impact on morale. "For the past two years the opening of college has been attended by an air of depression that has hung like a cloud over us through the year," observed the editors of the *Penn State Engineer* in October 1927. "This year there was none of that—instead the air seemed full of sunshine, hope, and promises of a great year in Penn State's history." This mood of optimism prevailed into 1928, as workers began moving earth and laying brick for the building that was to be the School of Engineering's new home. Not even the defeat of the College's proposed bond issue at the hands of Pennsylvania's voters could dampen the buoyant spirits of the engineering students and faculty. The bond question had finally cleared all legislative hurdles and had been placed on the ballot at the general election in 1928. There it failed by a mere 26,000 votes from a total of over 1,000,000 votes cast. Governor Fisher headed off the shadow of despair that would have inevitably settled over the College by pledging to use current and anticipated surpluses in the treasury for many of the same projects that were to be financed by the bonds. In May 1929, he signed a bill allocating Penn State $6.3 million for the 1929–31 biennium—the largest appropriation in the institution's 74-year history. Included in this amount was a record $2.25 million for new buildings.

But even if the Governor had not come to the rescue of the College after the rejection of the bond issue, the work of the School of Engineering would not have been seriously impaired. The new Main Engineering Building was already under construction and would be ready for use by the fall of 1929. None of the building funds appropriated by the Commonwealth were slated for any further physical expansion of the School. It did use some of its share of the general appropriation to purchase much-needed laboratory equipment and to make repairs to equipment and buildings that had seen only the skimpiest maintenance since the war, but on the whole, the addition of the new building had solved the most pressing problems of the School, which in most respects was in its best condition since the days of Louis Reber.

Just as in the waning days of World War I, prior to the disastrous fire, Dean Sackett had every reason to be sanguine about the future. Nearly all of the graduating seniors in 1929 had at least two or three job offers before commencement, as more companies requested more graduates from engineering schools across the nation and the field of occupations broadened. Employment trends gave no sign that this happy circumstance would change significantly. In addition to a strong demand for engineers, Sackett believed that the next few years were likely to witness the emergence of The Pennsylvania State College as a leading center of engineering research. Studies of heat transfer in building materials had already made the College preeminent in this field, while Paul Schweitzer was rapidly gaining recognition for Penn State in the area of diesel engineering, or more specifically, fuel oil spray in diesel engines. Schweitzer and his colleague, Kahlman J. DeJuhasz, were interested in perfecting fuel oil spray as a means of assisting manufacturers to produce more powerful and efficient machines. In

1927, after two years of strenuous, trial-and-error work, they constructed a pressure chamber—the only one of its kind in the world at the time—that enabled them to observe and measure the velocity and distribution of the oil spray that was injected into the cylinder prior to ignition. (Ignition itself was not yet a subject of study.)

A component of the Experiment Station, the diesel laboratory was the most advanced in the United States and compared favorably to similar laboratories in Europe. Impressed with the quality of the work being done there, the United States Navy arranged with the College early in 1929 to have a select group of officers begin graduate studies in diesel engineering. In June of that year seven naval officers arrived on the campus to begin a year of laboratory and classroom work that would ultimately lead to the awarding of a Master of Science degree. Each of the new students had at least several years of experience in submarine service, the Navy's principal area of diesel engine applications. When all seven successfully completed their studies in June 1930, a second contingent of eight officers replaced them. Penn State was one of only two schools that the Navy had invited to participate in its diesel engineering training program.

The staff of the Experiment Station did not have a monopoly on research. Several members of the Department of Mechanical Engineering were conducting studies in the areas of bearing design, journal friction, and lubrication of internal combustion engines, and faculty in the Department of Civil Engineering undertook extensive investigations aimed at improving the treatment of sewerage. In recognition of these and other research activities conducted under the auspices of the various departments, the School of Engineering had become by the close of the 1920s the site of numerous technical meetings and conventions. Some of the larger ones that met annually at Penn State included the Pennsylvania Water Works Operators Association and the Pennsylvania Sewerage Works Association (sponsored by the Department of Civil Engineering); the Pennsylvania Industrial Conference (sponsored by Engineering Extension); and the Industrial Organization and Management Conference (sponsored by the Department of Industrial Engineering). The newly formed (1927) Oil and Gas Power Division of the American Society of Mechanical Engineers held its first three annual meetings and six of its first ten meetings at Penn State, an unmistakable tribute to the School's fuel oil spray research program. In similar recognition of the nature of the work being done at the thermal laboratories, the American Society of Refrigerating Engineers chose the campus as the location for several of its yearly gatherings. In fact, in 1929 Professor A.J. Wood was elected president of the ASRE,

Dean Sackett and his department heads, 1928. *From left to right, front:* Breneman, Walker, Sackett, Kinsloe, Wood; *rear:* Hechler, Beese, C.L. Harris, Keller. (*La Vie*)

the first time in its 29-year history that an individual on the faculty of an engineering school held that office.

The increasing demand by private industry for the consulting services of the engineering faculty provided further evidence that the School had thrown off the malaise that had plagued it earlier in the decade. Illustrative of the diversity of consulting work that occupied faculty in 1929 were the activities of A. J. Wood, whom the Dry Ice Corporation of America had engaged to study the insulating properties of carbon dioxide; C. W. Beese, who conducted time and motion studies for the Timken Roller Bearing Company; Professor of Industrial Engineering Clarence E. Bullinger, who assisted in designing steel automobile bodies for the Budd Company of Philadelphia; and Professor of Electrical Engineering Charles Govier, whom the American Telephone and Telegraph Company hired to aid in the implementation of long-distance telephone lines. By all indications, the School of Engineering by the end of the 1920s appeared ready and able to assume a renewed importance in the affairs of the College, the Commonwealth, and the nation.

4 Depression and War: 1929–45

When the New York stock market suffered its fateful plunge in October 1929, few people anticipated the magnitude or duration of the subsequent decline of the American economy. Initially, what was eventually to become the Great Depression had a barely noticeable impact on Penn State's School of Engineering. As late as fall semester of 1931, 1170 undergraduates were enrolled in the School, down only slightly from the record enrollment of 1194 in 1927–28. The shrinking job market ultimately took its toll, of course. Dean Sackett somberly noted in his annual report for 1930–31 that only 50 percent of the School's 194 graduating seniors had received offers of employment prior to commencement. A year later, as economic conditions worsened, only 25 percent of the class of 1932 were able to secure jobs before or immediately after graduation. The nadir of the Depression occurred in June 1933, when just twelve seniors, or about 6 percent of the School's graduating class, had jobs waiting for them after commencement, and even that number Sackett judged "a good showing" considering the state of the economy. To make matters worse, chances were slim that students who failed to find employment in engineering positions soon after graduating could do so in the foreseeable future. A survey compiled by the School in the spring of 1933 revealed that only half the class of 1931 and barely a fourth of the class of 1932 were working as engineers.

The distressing employment outlook for young engineers made itself felt most obviously in the smaller classes of freshmen entering the School during the early 1930s. By the time the new academic year began

Table 1. UNDERGRADUATE ENROLLMENT

	1929–30	1933–34
Architecture	75	49
Architectural Engineering	80	51
Civil Engineering	214	113
Sanitary Engineering	26	11
Electrical Engineering	368	260
Electrochemical Engineering	41	53
Industrial Engineering	110	104
Mechanical Engineering	222	252
Railway Mechanical Engineering	2	0
Milling Engineering	2	1
Total	1140	894

in the fall of 1935, the effect of the Depression could be seen on all four classes. Total undergraduate enrollment stood at just 854 students—the lowest number in nearly 20 years. Many prospective engineering students who desired to attend college simply could not afford to do so in those troubled times. The total cost of a year at Penn State ranged between $600 and $800 during the 1930s. The very likely possibility of being unemployed after enduring four arduous years of preparation probably caused more students to turn away from engineering than did the cost of a college education, for total undergraduate enrollment at Penn State climbed steadily throughout the Depression. Whereas engineering students comprised about a third of the total undergraduate population of 3780 in 1927–28, they represented less than a fifth of a student body of 4982 in 1935–36. Engineering during this same period had fallen from first place in enrollment among the College's seven undergraduate schools to third, behind Liberal Arts and Agriculture. This slide was hardly unique, as engineering education bore the brunt of the Depression on most of the nation's college and university campuses.

Enrollment trends in the individual curriculums, as shown in Table 1, mirrored another phenomenon that was essentially nationwide. The Departments of Architecture and Civil Engineering were especially hard hit by the enrollment decline. The federal government's public works projects displaced many consulting engineers in these two fields, and the pace of privately financed construction slowed drastically. The demand for electrical engineers likewise fell, as the electrical manufacturers cut back production. By contrast, with business searching for ways to increase efficiency and lower production costs, the Department of Industrial Engineering was able to hold its own. The Department of Mechanical Engineering actually gained students, and for the first time in the twentieth century became the School's largest department. This growth, attributable mainly to industry's need to maintain a physical plant already in existence (as opposed to new construction or expansion), came in spite of the elimination of the railway mechanical engineering and milling engineering curriculums. Owing to their highly specialized nature, these curriculums had never attracted many students and were simply unable to weather the forced economies of the Depression.

The Depression's malevolent influences were not restricted to a decline in the number of students. Beginning in 1931, all departments cancelled their annual inspection trips, by order of the dean. The students themselves had always paid most of their own expenses for these four- to five-day trips, but, as one upper classman commented in the *Penn State Engineer* in January 1931, "the average senior in the Engineering School is not financially able to make the trip without undue and unfair drain on his pocketbook." And while most students traditionally considered these outings to be the highlight of their four years at Penn State, few found themselves in such comfortable circumstances that they could question the wisdom of Sackett's decision.

The economic downswing brought hardships to the *Penn State Engineer*, too. That journal, which had become a monthly (October-May) publication in 1927, gradually reduced the number of pages it carried after 1930, as advertising space became exceedingly difficult to sell to the large industrial and engineering firms that had previously supported the periodical so generously. In October 1936 the incoming editorial staff voted to cease publishing altogether, pending a thorough evaluation of the periodical's finances. After concluding that income from national advertising was no longer adequate to underwrite most of the costs of the *Penn State Engineer* in its earlier format, the editors severed their relations with the Engineering College Magazines Association and in February 1937 issued the first in a series of revamped *Engineers*. The new journal was little more than a mimeographed newsletter, distributed free, and carrying a smattering of local ads. Coverage was confined to campus affairs. As one of the editors finally confessed, the *Penn State Engineer* in this format was "a mere shadow of a once-successful and widely read store of technical articles," and was not at all worthy of a first-rate school of engineering. Still, the publication survived, publishing on a regular basis, which was more than could be said of the publications of many other engineering schools. At

Paul Schweitzer in the diesel laboratory. (Penn State Collection)

length, in February 1938, the *Penn State Engineer* was able to return to a full-size, slick-paper format. Gradually the number of articles and news items it carried pertaining to national engineering issues and events began to increase. Price was set at ten cents per copy. For the next twenty-three years it continued to appear regularly during the academic year, even during the paper and manpower shortages of World War II. In October 1961 the *Penn State Engineer* was superseded by the *Spectrum,* a more comprehensive scientific and technical journal published jointly by students in the Colleges of Engineering and Architecture, Mineral Industries, and Chemistry and Physics. In 1969 the journal again became the exclusive publication of engineering students and was renamed the *Penn State Engineer.*

Progress Amid Adversity

Not all activities of the School of Engineering were as seriously affected by the Depression as the *Penn State Engineer.* Significant advances were made in engineering research. Professors Schweitzer and DeJuhasz and their associates in the diesel laboratory widened their investigations to include ignition as well as injection problems. The laboratory itself continued to be the only facility of its kind in the United States. Meanwhile Professor Harold A. Everett of the Department of Mechanical Engineering supervised inquiries directed toward developing improved lubrication systems for gasoline and diesel engines. Heat transfer and sewerage treatment, both topics of longstanding concern to mechanical and civil engineering faculty, along with community planning, a subject of great interest among members of the Department of Architecture, also ranked high on the list of principal fields for research. Most of the forty or so individual research projects undertaken annually during the early and mid-1930s were very limited in scope, however, and often involved only a single faculty member working on his own time. The School's yearly research budget averaged about $25,000 during these years. The sum was several thousand dollars larger than in previous years and most welcome in light of economic conditions, but it still was not large enough to support more than a handful of major studies.

The School derived much of its research budget—in some years over half of it—from external sources, namely, private firms and industrial groups. The Depression had forced many companies to cut back or eliminate their own research facilities, increasing their dependence on the resources of colleges and universities. The Pennsylvania Grade Crude Oil Association, various diesel engine manufacturers, Texaco, and the Chemical Foundation of New York were among the more noteworthy supporters of engineering research. The money furnished by these and other private organizations gratified Dean Sackett and the faculty, since they permitted the School to continue studies which otherwise would have been curtailed for the duration of the Depression. President Hetzel, too, was pleased, yet the external research monies received by the Schools of Engineering, Chemistry and Physics, and Mineral Industries (formerly the School of Mines) at the same time caused him deep concern. In a March 13, 1933, memorandum to the College's Council on Research, which consisted of representatives of all the major academic divisions of the College, he outlined the reasons for his uneasiness. "There is a danger that applied research in cooperation with industries may overshadow and stifle institutional research of a more fundamental nature," Hetzel stated. "The question is not whether such investigations should be eliminated, but how far can we go in this line of work without prejudicing the instructional and research programs of an institution that is supported by public funds and unrestricted by contract."

At the request of the President, the Council adopted a general policy for the College which stated that cooperative research should remain a "minor feature" of each school's research program. Moreover, no school was to undertake any privately funded research that did not contribute to the overall institutional research program and that did not promise to yield results of public value, as contrasted with value primarily to the donor. President Hetzel perceived a disparity between this stated policy and actual practice, and in 1936 he asked the Council on Research to allow the administration to scrutinize the cooperative research projects of each school on an annual basis to ensure that these activities were in compliance with College policy. Dean Sackett opposed Hetzel's proposal. He termed the President's fear that research being conducted by the School of Engineering might somehow compromise the public character of the College "an exaggeration." The School of Engineering, Sackett pointed out in a letter to President Hetzel, had consistently followed a publicly oriented policy with regard to any research it accepted. No work was done unless the donor clearly stipulated that the School might make public any and all results and could, in the name of the College, have the right of patent, if applicable. Sackett also reminded Hetzel that the School of Engineering never consented to conduct research for a donor if that research could be done in the donor's own laboratory.[1] His appeal, along with those of the other deans, had its intended effect, and the President withdrew his proposal for review power.

If the nature of cooperative research occasionally troubled Hetzel, he harbored no qualms about the off-campus professional work of the engineering faculty, which during the 1930s became heavily public service-oriented. To cite only a few examples, both Professors Clinton L. Harris and Lewis Pilcher of the Department of Architecture secured leaves of absence to work for the federal government. Harris's investigation of wind pressures on tall buildings for the National Bureau of Standards attracted considerable popular attention in newspapers and magazines as well as professional acclaim in the technical press. Pilcher worked for the Public Works Administration as a consultant on public building design. J. E. Kaulfuss, Professor of Civil Engineering, served as a technical advisor to the Pennsylvania Department of Highways for a year, as that agency geared up for a major program of road construction. Assistant Professor of Industrial Engineering Amos E. Neyhart, while doing research for a master's degree in psychology, took note of the disturbing maladjustment between man and his most beloved machine, the automobile. People simply could not or would not learn the techniques of safe driving. Suspecting that the fault lay in improper training, Neyhart instituted on his own time and at his own expense the first high school driver education classes in the country at the State College High School in 1934. The response from parents and educators so overwhelmed him that in the summer of 1936 Neyhart administered the nation's first teacher preparation classes in driver training at Penn

State. He then took a two-year leave of absence from his department to organize a nationwide driver training program under the auspices of the American Automobile Association. Neyhart's official connection with the School of Engineering ended in 1938 when he was appointed director of the College's new Institute of Public Safety, a post he held for the next twenty-six years, during which time millions of persons learned to drive using the methods he perfected. Another member of the faculty who invested much of his own time and money in research and instruction was Eugene Woodruff. One of the few teachers in the School of Engineering to hold a doctorate, he came from a wealthy Michigan family. A millionaire, Woodruff pursued a low-paying career as an engineering professor because he found it personally satisfying. He refused to accept more than an instructor's salary, and even that, he liked to remark, could not begin to pay his income tax.[2] Nor was he in engineering for professional acclaim. Despite his extensive investigations in the fields of electric railways and later radio, he authored few papers based on his work. He delighted in helping students, however, and often held his radio engineering classes in his home, where students could make use of the short wave radio station he had installed in the rooms above his garage. The equipment there was far more elaborate and up-to-date than the College's own facilities.

The sharp decline in undergraduate enrollment during the 1930s allowed the teaching staff to devote more time not only to research but to the supervision of graduate studies. Dean Sackett had always regarded the lack of a diversified graduate curriculum to be his school's most glaring weakness. Until 1929 and the arrival of the Navy's submariners, the number of graduate students in residence rarely exceeded a dozen in any year. "Our opportunities to give graduate instruction are much restricted by two facts," Sackett noted in his annual report for 1928–29. "First, we do not have the funds with which to attract graduate students as Illinois, Purdue, and other institutions are doing; second, our teaching loads are still so heavy that little time or energy is left to conduct research." The Depression had in large part eliminated the heavy teaching loads. Sufficient funds with which to lure potential graduate students were another matter, however. Throughout most of the 1930s, only three sources of graduate fellowships existed: the Elliot Company, Texaco, and the Oil and Gas Division of the ASME. Of these, only the Elliot Fellowship in Research Engineering provided dependable support. It had been established in 1921 by W. S. Elliot, president of the Pittsburgh-based Elliot Company, a leading manufacturer of electrical equipment and an employer of many Penn State engineering graduates. The Texaco Fellowship, a more recent award, often had to be shared with the School of Chemistry and Physics, while the Oil and Gas Fellowship was offered only intermittently. Each department did have a few graduate assistantships to award annually, but these carried only a very small stipend and stipulated that the graduate student must teach part-time, which correspondingly lengthened the period of his studies. Nevertheless, many seniors opted to begin graduate studies rather than face the vagaries of an uncertain job market. As small as the remuneration might be, at least it provided an alternative to the bread line and kept the student in his chosen profession.

Graduate enrollment rose slowly but inexorably, from about a dozen students in residence in 1927–28 to 25 in 1935–36. Mere numbers did not reveal an even more important change occurring in the engineering graduate student body. In previous years, many of the persons earning advanced degrees were the School's own faculty members, who often came to the College with only a bachelor's degree plus several years of work experience. Now, many graduate students came to Penn State immediately after completing their undergraduate education, sometimes from such prestigious institutions as the Massachusetts Institute of Technology, Cornell University, and Rensselaer Polytechnic Institute, although students with undergraduate degrees from Penn State still predominated. The majority of graduate students elected to pursue studies in fields in which they could utilize either the thermal laboratories or the diesel engineering laboratory, since these were the School's most modern and complete research facilities. All candidates for advanced degrees except those in architecture came under the immediate administrative jurisdiction of the Engineering Experiment Station rather than the individual departments. Following the creation of the Graduate School in 1922, the degree most commonly awarded graduate engineering students was that of Master of Science. The older technical degrees that had traditionally placed more emphasis on the candidate's practical experience were conferred with decreasing frequency and were discontinued entirely by 1955. The awarding of the first doctoral degree in engineering came in 1936, when Theodore B. Hetzel (no relation to President Ralph Hetzel) received a Ph.D. in mechanical engineering. Hetzel did his thesis research in the field of oil sprays, under the supervision of Professor Schweitzer.

Had the size of the faculty declined in proportion to the precipitous drop in undergraduate enrollment, conditions for graduate study would not have been so favorable. Happily, Dean Sackett was able to retain most of the teaching staff. The high turnover of the 1920s had practically disappeared, with the faculty of 1928 still forming the core of the instructional staff a decade later. This stability resulted not from any substantial increase in salaries—Penn State still remained far down the list in this regard—but from a dearth of attractive opportunities elsewhere. Sackett no longer had to worry about losing his most promising teachers to the more lucrative callings of private industry. Also contributing to faculty stability were the reduced size of many classes and the increased time available for research, both of which made conditions much more conducive to the retention of a first-rate faculty.

Two changes did occur in departmental headships, however, that were to have important consequences for the future. In 1930 Charles Beese resigned as head of the Department of Industrial Engineering in order to take a job with the Armstrong Cork Company. He later became a prominent industrial engineering educator at Purdue University. Beese was succeeded by Clarence E. Bullinger. A native of Philadelphia, Bullinger had joined Penn State's industrial engineering faculty in 1922. For several years prior to that, he simultaneously held positions as machinist, draftsman, and student at Philadelphia's Drexel Institute of Technology. In 1931, in an appointment surrounded by more tragic circumstances, Harold A. Everett succeeded A. J. Wood as head of the Department of Mechanical Engineering. Wood died in 1931 as a result of injuries received when struck by a motorcycle while crossing the street near his home. He had made valuable contributions to the School of Engineering and to the engineering profession through his work at the thermal laboratories and in the railway mechanical engineering course. His skills as a teacher and administrator would be missed. Everett was no less qualified, however. A 1902 graduate of the Massachusetts Institute of Technology, he had taught at both MIT and at the Naval Academy and had held supervisory engineering positions with a shipbuilding firm before joining the Penn State mechanical engineering faculty in 1922. Turbine propulsion especially interested Everett, and he had done a considerable amount of research and consulting work in that field before shifting his attention in the 1920s to lubrication problems of internal combustion engines.

Besides graduate work, another element of the School of Engineering that actually grew stronger in some ways during the Depression was the Department of Engineering Extension. In the late 1920s, extension was divided into two segments, the correspondence division and the class division, headed by Hugh G. Pyle and Irving C. "Hix" Boerlin, respectively, while J.O. Keller continued to supervise overall operations. The correspondence division handled all courses that were taken by mail. The class division encompassed the six evening branch centers (at Allentown, Wilkes-Barre, Scranton, Reading, Williamsport, and Erie), the class centers, and miscellaneous independently sponsored sessions, such as foreman training classes. By 1929 both divisions had given instruction to over 30,000 Pennsylvanians on virtually every current engineering topic.

The correspondence division did not bear up well under the brunt of the Depression. The number of students taking courses by mail fell from 1293 in 1929 to 466 in 1934. Much of this decline resulted as firms that had previously underwritten the costs of correspondence lessons for their employees withdrew this support. The class division, too, suffered from the inability to supply the financial backing that the foreman training and similar courses depended upon. Only the very largest companies, mainly those in the aluminum, steel, textile, and pulp and paper industries, could afford to continue with their extension programs. By contrast, enrollment at the class centers and the evening branch centers burgeoned, with the exception of the evening school at Williamsport, which was discontinued for lack of interest in 1930. Many students who could not afford to spend four years in college discovered that they were still able to obtain an education in many phases of engineering by attending these extension classes on a part-time basis. The popularity of the classes soared despite the fact that none of the two- and three-year certificate programs offered by the branch centers carried college credit. Enrollment in the class division climbed from 1862 in 1929 to 3122 in 1934.

This same growth characterized the extension programs of most of the College's other schools as well. The time had therefore arrived, President Hetzel believed, to supply unified administrative supervision to nearly all of the College's extension activities. To oversee this new Division of Central Extension, Hetzel selected J. Orvis Keller, who in 1934 assumed the title of Assistant to the President in Charge of Extension. Keller was a logical choice. A nationally recognized figure in the field of technical extension education, he had done a superlative job in heading the Department of Engineering Extension since 1925. Unlike the concept of centralized extension that had been considered during President Thomas's administration, Hetzel's plan allowed the extension programs of the Schools of Chemistry and Physics, Education, Engineering, Liberal Arts, and Mineral Industries to retain most of their autonomy. The School of Agriculture was not included in the new extension division because of the unique federal-state sponsorship of its extension arm and the unyielding opposition by agricultural interests to any attempt to compromise the independence of the extension service. Keller and his small staff were primarily responsible for coordinating the extension activities of the various schools and handling the bulk of the routine administrative chores.

In addition, Keller directed the establishment in 1934–35 of the first four of Penn State's branch campuses, or "undergraduate centers." The College set up these centers—the foundation of the Commonwealth Campus system of later years—in response to the public demand for full-time, collegiate-grade instruction for students who could not afford to attend the main campus. The centers offered basic general education classes for freshmen and sophomores of all majors, including engineering. They were not intended to compete with the School of Engineering's five evening branch centers, which continued their terminal certificate programs. At Dean Sackett's recommendation, President Hetzel named Edward L. Keller to succeed J.O. Keller as head of the Department of Engineering Extension (the two Kellers were not related). Edward Keller had received a bachelor's degree in industrial engineering from Penn State in 1925 and was already well versed in the aims and operation of the department, having served as J.O. Keller's assistant for the past several years.

By the time the new academic year began in September 1934, the most critical stage of the Depression had passed. Several more years of economic hardship lay ahead, but the worst was over. The freshman class entering the School of Engineering in 1934 was 10 percent larger than the previous year and marked an end to four consecutive years of decline. Employment prospects were also beginning to brighten. Nearly all the class of 1935 secured engineering positions within two months after commencement in what was the best job market for engineers since 1930.

Dean Sackett could take pride in the fact that, in the face of an unprecedented drop in undergraduate enrollment, engineering at Penn State experienced growth in certain areas and emerged from the depths of the Depression healthier than at many other land-grant institutions. Conclusive evidence of the integrity of the instructional program came in 1936, when the Engineers' Council for Professional Development evaluated the curriculum of the School of Engineering. The ECPD had been created in 1932 by five professional engineering societies, the Society for the Promotion of Engineering Education, and the National Council of State Boards of Engineering Examiners in response to a rising professional consciousness among engineers, about 75 percent of whom held college degrees. Over 135 colleges and universities, varying greatly in their caliber of instruction and the adequacy of their facilities, offered engineering degrees in one or more fields in 1932. One of the ECPD's chief goals was to develop standards by which the quality of engineering education could be measured, standards which then could form the basis for admission to the professional societies and the registration and licensing of engineers. More and more states were moving to license engineers, but unless a uniform basis for evaluation could be achieved, each state was likely to utilize different standards and practices, resulting in mass confusion nationally. To the ECPD's Committee on Engineering Schools fell the tasks of formulating guidelines for accrediting engineering curriculums and conducting field evaluations. By the end of 1937, evaluating teams had visited 129 schools having a total of 626 degree curriculums in engineering. Based on the findings of these teams, the ECPD granted full, five-year accreditation to fewer than 60 percent of these curriculums.[3] Penn State's School of Engineering was one of only a handful of institutions in New England and the Middle Atlantic states that received accreditation for every one of the curriculums the ECPD had examined. Dean Sackett hailed this feat as "one of the outstanding achievements of the last ten years" and "evidence of the standing of the School among those who know."[4]

Certainly no better legacy could be left to his successor, as Sackett in 1937 reached age seventy and announced his impending retirement. An engineering educator of the first rank, he had devoted the last twenty-two years to the improvement of engineering education not only at Penn State but also, through his participation in the Society for the Promotion of Engineering Education, throughout the nation. His efforts would win for him in 1938 the Lamme Award, SPEE's highest honor. Yet Sackett found time for other affairs, too, and was an individual of remarkably varied accomplishments. A painter of no little talent, he exhibited his water colors in several one-man shows in the Northeast. A firm believer in the character-building potential of intercollegiate athletics, he served as chairman of Penn State's Faculty Senate Committee on Athletics and for five years was vice-president of the National Collegiate Athletic Association. He consistently took an uncompromising stand for honesty and integrity in one of intercollegiate sports' most troubled eras. That same commitment to character development was evidenced in his efforts as a founder in 1928 of the Penn State chapter of Triangle, an international fraternity dedicated to scholarship and fellowship among engineering students, begun at the University of Illinois in 1926. And somehow, amid all his other concerns, the dean found ample time to engage in his favorite boyhood pleasure of sailboating and became an experienced navigator on the Great Lakes and along Long Island Sound.

In appreciation of these numerous professional and personal contributions to the College, Penn State's board of trustees bestowed the rank of Dean Emeritus of Engineering upon Sackett, the first recipient

Harry P. Hammond (Penn State Collection)

of this honorary title. After his retirement, he made his home in New York City, although he continued to maintain an active interest in Penn State affairs and in the activities of the Engineers' Council for Professional Development. He died in New York in 1946. Robert L. Sackett made no claim to being a bold innovator or a great visionary. Such qualities would have been of limited value, anyway, in those difficult times following the disastrous fire of 1918 and the onset of the Great Depression in 1929. He did possess unsurpassed administrative abilities and a stubborn unwillingness to let adversity deflect him from his chosen course, attributes that were much more attuned to the needs of his era.

Harry Hammond: Entering a New Epoch

The new dean of the School of Engineerng possessed academic credentials no less impressive than those of his predecessor. Fifty-three-year-old Harry Parker Hammond, a native of Asbury Park, New Jersey, had received his undergraduate degree in civil engineering from the University of Pennsylvania in 1909. He then taught in the civil engineering department at his alma mater while working toward the advanced degree of Civil Engineer. Upon completing his graduate studies in 1911, Hammond accepted an appointment as an instructor in civil engineering at Lehigh University. A year later he joined the civil engineering faculty at the Brooklyn Polytechnic Institute, remaining at that institution until coming to Penn State in the summer of 1937. While at Brooklyn, Hammond also did a limited amount of consulting work in his field of specialization, sanitary engineering, for various governmental bodies and private firms. Harry Hammond's real interest lay in engineering education, an area in which his qualifications were exceptional. Throughout his tenure at Brooklyn, where he eventually became head of the civil engineering department, Hammond took an active role in the affairs of the Society for the Promotion of Engineering Education. In addition to chairing SPEE's civil engineering section for several years, he served during the 1920s as associate director of the famous Wickenden Commission. This panel, another offshoot of SPEE, conducted a comprehensive survey of engineering education and made a number of important suggestions for its improvement. In 1934, Hammond, as a representative of SPEE, helped found the Engineers' Council for Professional Development and participated for two years as a member of the ECPD's evaluating team. In 1936 he was elected president of the Society for Promotion of Engineering Education. President Hetzel and the engineering faculty considered themselves extremely fortunate to have an individual of such national stature take over the direction of Penn State's School of Engineering.

Hammond was impressed by what he termed "a deep and abiding, one might almost say fierce, spirit of loyalty to the College and to the School" on the part of faculty and students. In welcoming new and returning students at the beginning of the new academic year in 1937, the new dean pledged to "preserve the traditional values that have been produced by [my] predecessors and their associates," as well as to "place Penn State still further forward among the leaders of engineering and architectural education."[5] Hammond intended these words to be more than the typically platitudinous oratory one might expect on such an occasion. His broad experience as an educator had convinced him that certain changes and improvements were long overdue in the Engineering School, and he expected to devote the bulk of his attention to ensuring that they were implemented.

One of the improvements most urgently required was an expansion of the School's physical plant. Insufficient building space had not troubled Dean Sackett much after 1929, for the construction of the new Main Engineering Building and the subsequent drop in enrollment during the Depression combined to make available more room than the School had ever had previously. As economic conditions improved and the student body began to grow again, lack of space once more became a problem. Sackett had foreseen the difficulties that would confront his successor. In his final report to President Hetzel, the departing dean ranked the acquisition of more building space as far and away the School's most pressing need. Specifically, he proposed the erection of a new electrical engineering building and the enlargement of the Main Engineering Building to the full size originally envisioned. The Department of Electrical Engineering was then housed in the cramped quarters of Units D and E; its laboratories were small and its equipment outmoded. Putting this department in a building of its own, with completely new furnishings, would release space in the older units for use by other departments. Expansion of the main building, said Sackett, would give the School the auditorium and the library that it had needed for so many years, as well as provide more space for the Department of Architecture, most of whose fine arts classrooms were located in the badly deteriorating Unit F.

Perhaps realizing that obtaining funds for these two projects would be a formidable enough task, Sackett had made no recommendations for expanding the facilities allotted for engineering research. Nevertheless, the Experiment Station also suffered severely from insufficient room. The station used a small portion of Unit F for offices, where they were in close proximity to the thermal laboratories, which were rapidly outgrowing the original installation of 1911. The Mechanical Engineering Laboratory housed the diesel laboratory until 1931. In that year the College's new power plant went into operation, and the diesel laboratory was moved to the old power plant that had once adjoined the old Main Engineering Building. (To make way for the new plant, the old Mechanic Arts Building and its appendages were demolished.) The diesel laboratory had to share its new home with the School of Chemistry and Physics' new Petroleum Refining Laboratory, so space remained very much at a premium.

When Harry Hammond became dean, he vigorously seconded Sackett's call for a new electrical engineering building. In fact, President Hetzel had already included the proposed structure in the multimillion dollar building program he had recently prepared. The Penn State administration had formulated this program upon the creation of the General State Authority by the Pennsylvania legislature in April 1937. Through a complicated series of agreements with the federal government's Public Works Administration and the College, the GSA underwrote most of the cost of the College's construction projects, and the PWA provided an additional $1 million. In February 1938 President Hetzel and Governor George H. Earle presided at a joint groundbreaking ceremony for the new electrical engineering building and a dozen other structures, including a library, new additions to the Mineral Industries and Liberal Arts buildings, and new buildings for the Schools of Chemistry and Physics, Agriculture, and Education.

The new home for electrical engineering was situated behind the Mineral Industries Building, with the portico of the main entrance fronting on the President's House. Designed by GSA architects Hunter and Caldwell of Altoona, the building was composed of brick and limestone to harmonize with nearby structures and contained three full floors. Following past practice, plans called for the construction to proceed in two phases. A T-shaped structure was erected immediately. Later, when funds permitted, another 200' × 50' wing was to be added, giving the building an H-shaped layout. Although plans called for the building to be ready for use within two years of the start of construction, the general contractors, McCloskey and Company of Philadelphia, encountered limestone caverns when excavating the foundation. This circumstance resulted in a slight delay, so that the Department of Electrical Engineering could not take full possession until the summer of 1940. On the first floor were located classrooms, a large assembly room, offices, and the dynamo, illumination, and standards laboratories. The second floor contained mostly classrooms and smaller seminar rooms. An acoustics lab and laboratories for electronics and communications were situated on the third floor. Pending the completion of another wing, the high voltage, electric traction, and electrochemistry laboratories remained in Unit E.

The construction of the Electrical Engineering Building did not, in Hammond's estimation, fully satisfy the space requirements of the School. Rather than press for an expansion of the Main Engineering Building, the dean believed that higher priority should be given to enlarging the Mechanical Engineering Laboratory. He wrote to President Hetzel on this subject in May 1938. "It may seem a strange time to speak of problems of space congestion when the College is in the midst of a great building program, yet no relief for the condition in the mechanical engineering lab and no new research work can be afforded, and no new lines of work undertaken, until the originally planned wings for this building are erected."[6] He pointed out that the Mechanical Engineering Laboratory was built in 1920, when 260 students were enrolled in the department. Now the department had to squeeze over 390 undergraduates into the same space and still provide room for graduate students and faculty research projects. Unfortunately, Hetzel could do little to assist his colleague in relieving the overcrowded conditions, since the GSA and PWA monies had already been allocated and no further building appropriations were forthcoming.

Civil engineering students learning to adjust transits during summer camp at Stone Valley. (Penn State Collection)

Hammond's attempts to improve the facilities of the Department of Civil Engineering were more successful. The distance of the Lycoming County summer camp site from the College and the lack of a satisfactory body of water there for hydrographic studies had long been sources of displeasure to Elton Walker and other members of the department. To make matters worse, rental rates paid by Penn State for the use of the property rose sharply after 1929. Dean Sackett had therefore begun to search for a new location for the camp, one nearer the College and upon which some permanent structures could be erected. When Dean Hammond succeeded Sackett, negotiations were already in progress between the College and the United States Department of Agriculture regarding a large tract of land in Stone Valley, Huntingdon County, about 15 miles from the central campus. The Agriculture Department's Farm Security Administration acquired the land in the early 1930s as part of its scheme to relocate families from submarginal farm land. Late in 1938 an accord was reached whereby the Department of Agriculture leased 6000 acres in Stone Valley to the Pennsylvania Department of Forests and Waters, which in turn subleased for 99 years 4500 acres to Penn State to be used for practicums by the Departments of Civil Engineering and Forestry and various departments in the School of Mineral Industries. The federal government's Soil Conservation Service built an access road to the site and a small reservoir. The College supplied materials for several buildings, tent platforms, and sewerage facilities, all of which were constructed by the Soil Conservation Service labor. Total cost amounted to $36,000, of which Penn State contributed $6000. Construction of the civil engineering camp alone cost $17,000, of which the College paid $3800. The first student civil engineering party encamped at Stone Valley in 1939, although construction work was not finished until 1940. Beginning in the late 1950s, the Stone Valley facility was converted to a University recreation area, and summer camps for civil engineering students were eventually phased out.

A second area in which improvement, or more properly speaking, reform, was needed centered on the engineering curriculum. The Wickenden Report, issued in 1929 by a SPEE committee chaired by William E. Wickenden, had already suggested the direction in which these reforms should proceed.[7] To a large extent, the Wickenden study echoed the findings of the Mann Report, which had been overshadowed by World War I and had not received the attention it deserved. Undergraduate specialization should be de-emphasized, and the number of specialized curriculums should be reduced to a manageable number. (Some engineering schools offered bachelor's degrees in as many as thirty or forty separate fields.) Social studies, particularly English and economics, and science should form a larger part of the undergraduate curriculum; and along with engineering methods, they should constitute the three basic stems of engineering education. Among its other recommendations, the Wickenden committee urged that the length of the baccalaureate program be limited to four years and not, as practiced by a few schools, stretched to a fifth year to include more specialized studies. Specialization was deemed to be the proper objective of graduate education, and engineers were encouraged to continue their professional training in graduate school or through the education programs offered by their employers. Many institutions were reluctant to act on the conclusions of the Wickenden Report until forced to do so by the Depression. By then, engineering graduates were accepting jobs far removed from the specialized applications for which they had studied, making many educators reevaluate the advisability of offering highly specialized training to undergraduates. Many engineers chose advanced study in lieu of employment, making it imperative for many schools to reassess their graduate curriculum. Disillusioned by technology's inability to preserve prosperity, some engineering students themselves began calling for the addition of more non-technical subjects to the curriculum. "The engineering course is organized wholly around making a living, even though this is but a part of, and secondary to, the art of living," complained a student in a letter printed in the *Penn State Engineer*. "No attempt is made to provide even a sketchy background of historical and philosophical knowledge."

That student's concern notwithstanding, Penn State's School of Engineering since John Price Jackson's day had required a relatively large number of credits in the kinds of humanistic studies mentioned in the Wickenden Report. Dean Hammond recognized the value of a liberal education to engineers. "Preoccupation with the purely technical sides of engineering should not preclude an appreciation of the social and economic results of the things that engineers do," he wrote in his monthly column in the *Penn State Engineer* in April 1938. He admitted that some institutions did allow an avalanche of technical studies to bury a student's budding interest in social and cultural topics, but so far as he could determine, that was not happening at Penn State.

In other areas, too, the School of Engineering was already in compliance with the directives of the Wickenden committee. Robert Sackett had been asking for increased appropriations to expand graduate work long before the Wickenden Report was completed. And the school had never permitted the quest for specialization to reach the proportions then current at many other institutions. Dean Hammond did see a need for the addition of more mathematics and science courses to the curriculum, however. As a newcomer to Penn State and as an educator who was familiar with trends in engineering education at the national level, Hammond believed that in many cases the subject matter taught in the School had not kept pace with the explosion of technological knowledge that had occurred since World War I. Although he never admitted as much publicly, Hammond's private reports reveal that he sympathized with student dissatisfaction over what one undergraduate characterized as "the obsolete, misplaced, and uninspiring courses" that predominated in many departments.[8] Moreover, not only had the subject matter of many courses failed to change with the passing years; so, too, had teaching methods. Teachers still relied almost exclusively on the lecture and the textbook. Hammond found that the criticism of engineering education he had voiced earlier as a member of the ECPD's Committee on Engineering Schools held true all too frequently at Penn State. "Instead of throwing the responsibility on the [student] to a greater extent than in former years, I suspect that the tendency has been to take it away," he had told a SPEE audience in 1933. "I believe the weakest spot in our entire scheme of instruction is in lecturing in the classroom on material assigned to the student in a textbook and repeating in different words the same things the author has said—in a word, studying the student's lesson for him."[9]

In 1939 Dean Hammond directed the faculty of the School of Engineering to begin a one-year evaluation of the curriculums of all departments and submit suggestions for improvement. In the meantime, Hammond agreed to head a similar survey of the engineering curriculum on the national level sponsored by SPEE's Committee on Aims and Scope of Engineering Curricula. The Hammond Report, published in 1940, essentially reinforced the suggestions made earlier by the Wickenden Report regarding the necessity of the four-year undergraduate curriculum, the value of graduate work, and the need for greater emphasis on science and the humanities.[10] It was not by mere coincidence that the recommendations made by the faculty of Penn State's School of Engineering paralleled those of the Hammond Report. These reforms, as proposed by the School and approved by the College's faculty senate in the fall of 1940, caused no radical transformation of past practices, but they did give Penn State the honor of being the first engineering institution to act on the guidelines laid down by the latest national survey.

One of the most important changes was the reduction in the number of credits required for graduation from an average of 154 to a uniform number of 146 for all departments. In another modification designed to achieve greater interdepartmental uniformity, the freshman year was made identical for all students save those in architecture, architectural engineering, and electrochemical engineering. Each department also reduced the number of courses* it offered in a given core of instruction, while holding the total number of credits awarded for that core approximately the same. The Department of Civil Engineering, for example, which previously demanded that students take a sequence of six courses in structural engineering worth 14 credits, now consolidated the same subject matter into four courses carrying 15 credits. Overall, the number of courses in that department dropped from 76 to 52.

Non-engineering subjects, too, came under close scrutiny. Because the number of credits in the humanities (22–25) required of engineering students was slightly higher than the national norm and well above the average for land-grant institutions, no changes were made in this area. A few modifications were effected in science and mathematics courses, however. The total number of required credits was increased

*By this time, "course" had assumed its modern definition as a sequence of instruction lasting a single semester or term while "major" now generally referred to a student's degree curriculum.

and, in cooperation with the Schools of Liberal Arts and Chemistry and Physics, an effort was undertaken to make course content more relevant to the engineering student's professional studies. Thus, for example, the Department of Mathematics lessened the emphasis on analytic geometry and strengthened work in differential equations in its sequence of math courses for engineers. A final change eased restrictions on course selection in professional subject areas for juniors and seniors, permitting students to choose from a greater number of electives. The curricular reforms eased teaching loads, too. The streamlining of course sequences eliminated duplication of subject matter and resulted in more efficient utilization of instructional time and resources. Uniformity of instruction was enhanced, yet students enjoyed more freedom than ever in scheduling courses to suit their needs.

The faculty recommendations did not directly address the issue of obsolete course content and pedagogical methods. Dean Hammond knew that one of the best means of promoting a fresh approach to engineering education was to introduce "new blood" into the teaching staff. He was in an especially favorable position to do this, having had 16 faculty positions fall vacant during his first two years as dean, including the headships of two departments. To succeed the retiring Paul Breneman in 1938 as head of the Department of Mechanics and Materials of Construction, Hammond selected Dr. Rudolph K. Bernhard. Born in Germany and educated in his native land and in England, Bernhard had served for a decade as Chief Structural Engineer for the German State Railways before coming to the United States in the 1920s. He had acquired an international reputation as a pioneer in the dynamic testing of structures and an inventor of many instruments used in these tests. Hammond allowed Bernhard to rebuild completely the old materials testing laboratory and modify much of its equipment to serve a greater variety of purposes. In 1940 the dean won trustee approval to change the name of Bernhard's department to the Department of Engineering Mechanics. Hammond also took the opportunity to suggest that the name of the School be changed to the School of Engineering and Architecture in order to give more recognition to architectural education, "a division of work not included under the general term of engineering," but the trustees took no action on this request.[11] The other new department head was Dr. Frederic T. Mavis, who took over the leadership of the Department of Civil Engineering upon the retirement of Elton Walker in 1939. Mavis, formerly head of the civil engineering department at Iowa State College, had earned his undergraduate and graduate degrees at the University of Illinois and had spent several years in

Europe conducting theoretical studies in hydraulics under the sponsorship of the American Society of Civil Engineers.

In addition to physical expansion and curricular reform, progress was made in several other areas during the early years of the Hammond administration. The School of Engineering's first cooperative education program began in the fall of 1937 under an agreement between the Department of Industrial Engineering and the York (Pennsylvania) Ice Machinery Company. Students in the program (eight, initially) spent the spring semester of their first four years acquiring on-the-job experience at the York factory. The summer sessions, the fall semesters, and the entire fifth year were devoted to formal instruction at the Penn State campus. Although an innovation at Penn State, cooperative education had been a standard feature of some of the best engineering schools for many years, having been introduced originally by Dean Herman Schneider at the University of Cincinnati in 1906. By the late 1930s, about 20 schools had adopted some form of cooperative education scheme.[12] The conservative leadership of Dean Sackett was partially responsible for Penn State's tardiness in experimenting with cooperative education in engineering, but an equally important factor was the rural location of the College and the resulting lack of nearby heavy industry with which the School of Engineering could develop suitable partnerships.

Laboratory equipment was upgraded to the extent permitted by the School's limited finances during these years. In 1938 the Department of Civil Engineering established a soil mechanics laboratory to examine physical properties of soils under various types of stress. The following year a $5000 annex to the thermal laboratories was built in conjunction with a project the Experiment Station had undertaken for the Navy. The Bureau of Ships asked the station to develop materials that would provide maximum insulation in both arctic and tropical climates. The most important addition to the equipment of the Department of Mechanical Engineering was a wind tunnel, put into operation in 1940 in the Mechanical Engineering Laboratory. A medium-sized tunnel, it cost over $5000, had a 3' × 4' throat, and could produce a maximum wind velocity of 125 miles per hour. Dean Hammond hoped that the wind tunnel would ultimately lead to a baccalaureate curriculum in aeronautical engineering. As early as 1927, Harold Everett had taught several courses in aeronautical engineering as an informal option for senior mechanical engineering students, but despite strong student interest, little more had been done to expand instruction in this field. In the late 1930s the department did contract with State College airport operator Sherman Lutz to give undergraduates taking aeronau-

Interior view of the wind tunnel. (Penn State Collection)

tics courses ground and flight training. This training had limited appeal, however, since it carried no credit, and students were required to pay their own expenses.

One reason for the apparent indifference shown toward aeronautical engineering was Professor Everett's belief that the School of Engineering ought to direct its resources instead toward beginning a curriculum in naval architecture and marine engineering. Everett contended, perhaps predictably in view of his undergraduate training in marine engineering, that "the needs of the [aircraft] industry are amply served by the colleges presently giving four-year courses in aeronautical engineering," at a time when the demand for marine engineers and naval architects far exceeded the supply.[13] Hammond disagreed. In May 1941 he told President Hetzel that Everett frequently expounded on the advantages of a curriculum in marine engineering, but, said Hammond, "I have not encouraged the idea because I have not seen where the means of carrying out such a program could be found." The dean argued that "the need for aeronautical engineers is attested to both by student demand and by industrial need. There is no full-fledged curriculum in Pennsylvania in this field, though the needs are urgent and the opportunity evident. The Pennsylvania State College seems clearly to be the agency through which this need can best be met."[14]

In 1940 Hammond appealed to the General State Authority and the federal government for money to erect an aeronautical engineering building and to construct a college airport that could be utilized by the proposed aeronautical engineering department and for general aviation purposes. "As commercial and private flying increases in value," the Dean prophesied to President Hetzel, "the availability of landing facilities close at hand will become of considerable importance to the College and community." Both governmental agencies in the end reacted negatively to Hammond's request for funds. This setback did not prevent him from securing the appointment of Dr. David J. Peery as Professor of Aeronautical Engineering in the Department of Mechanical Engineering, nor did it prevent that department from adding a few more elective subjects in aeronautics to coincide with the opening of the wind tunnel.

Engineering and Another War

As the decade of the 1930s drew to a close, the School of Engineering was well on its way to a complete recovery from the ills that had befallen it during the Depression. Undergraduate enrollment stood at 1033 in 1939–40, ranking the School seventh in the nation. Approximately 120 engineering students, mainly freshmen, were attending classes at the undergraduate centers. Graduate enrollment totaled 27, placing Penn State ninth in this category. In the face of renewed growth of the student body and the various improvements made to the School, Dean Hammond still remained cautious in looking to the future. "It seems entirely unsafe to venture upon any quantitative prediction of future trends," he noted in his report for 1938–39. The hardships of the Depression still hovered dangerously near; another economic downswing was a very real possibility. War clouds were again gathering over Europe, just as they had done 25 years earlier, threatening to engulf civilization in another whirlwind of violence.

The uncertainties caused by the Depression could be trivial compared to those arising from a world war, a possibility suggested by events occurring in 1940 and early 1941. Germany had invaded Poland in September 1939, touching off declarations of war among the European powers and a campaign for military preparedness in America. Congress passed a conscription act in 1940 and, while engineering students and faculty held deferments, there was no assurance that they would enjoy this status indefinitely. Many members of the teaching staff belonged to the army or naval reserve and were subject to being called to active duty (as indeed a few were in 1940) with a minimum of advance notice. In the event the United States did become an active participant in the war, the burden of providing the technical training required by hundreds of thousands of men and women for war-related tasks could be expected to fall to the nation's colleges and universities, especially those having strong engineering and scientific schools. These circumstances, therefore, were hardly conducive to long-range planning in the School of Engineering. Consequently, Dean Hammond and the rest of the faculty had no alternative but to bide their time and await the march of events.

At length the situation began to resolve itself. In October 1940 the United States Office of Education appointed a special National Advisory Committee—one of whose members was Harry Hammond—to devise plans for instructing large groups of civilian students in defense-oriented technical subjects. The Engineering, Science, and Management Defense Training program (ESMDT) that grew out of the committee's work called for as many institutions of higher learning as possible to share in the responsibility of supplying this instruction. At Penn State, ESMDT got under way in January 1941. Since the courses were to be given on a statewide basis, their supervision came under the jurisdiction of the College's Central Extension office. The Department of Engineering Extension for all practical purposes was merged into Central Extension for the duration of the emergency, as were the extension departments of the other schools. By the end of the year, of the 14,000 individuals who had received training in one or more of Penn State's ESMDT courses, 10,000 of them had taken engineering subjects, representing the largest such enrollment at any institution in the country.[15]

The Office of Education also wished to institute more advanced courses in engineering that students who already possessed some technical training could take on college campuses. Many schools initially hesitated to participate in this phase of the ESMDT program, fearing the disruption that these sixteen-week courses would have on their peacetime routines. Penn State's School of Engineering, however, immediately pledged its cooperation. In January 1941 a course in production engineering began under the direction of Professor of Industrial Engineering Clarence Bullinger. The first class numbered thirty students—the maximum allowed by the government—all of whom had previously taken engineering courses at Penn State, although few had graduated. The Office of Education paid all tuition and other costs. Coincidentally with the inauguration of Bullinger's course, a similar sixteen-week course in diesel engineering began, supervised by Professor Everett. This course was initiated at the request of the Navy, which sent a detachment of thirty reserve officers for training. Reminiscent of the graduate courses in diesel engineering the School had provided for

Diesel engineering class for naval reserve officers. (Penn State Collection)

the Navy a decade earlier, the program was the first of its kind to be implemented under the Department of the Navy's new preparedness program. In April 1941 these two courses were completed with "excellent results," according to Dean Hammond. They were then repeated, with double the number of students and accompanied by a third course, this one in materials testing and taught by Professor Bernhard. Both civilian courses, each of which was taught for a third time beginning in October, aimed to prepare students to fill positions as inspectors, designers, and supervisors in the defense industries.

The bombing of Pearl Harbor on December 7, 1941, and America's subsequent entry into World War II invoked what in many respects was a repeat performance of the School of Engineering's role in the conflict of 1917–18. Early in 1942 the College's board of trustees voted to advance the date of commencement from June 8 to May 9. In a scheduling change of even greater proportions, the trustees acceded to a request from the National Advisory Committee of the ESMDT to place Penn State on an accelerated academic year in order to satisfy manpower requirements. The new year began in June 1942 and consisted of three full terms of approximately four months each. The class of 1943 thus received their diplomas in December 1942, and class commencements occurred triennially thereafter. Students could now complete a regular four-year program in less than three years.

Even before the war had begun, the School of Engineering was slowly losing its faculty, mainly reservists, to the armed forces. After Pearl Harbor, the losses grew alarmingly. During the first three

Assistant Dean of Engineering Royal Gerhardt (extreme left) and a group of Curtiss–Wright Cadettes. (Penn State Collection)

months of 1942, for instance, sixteen members of the teaching staff resigned or took leaves of absence, eleven to join the military and five to work for civilian government agencies. By September 1943, thirty-nine teachers, or one-third of the 1941–42 faculty, had left. The number rose even higher the following year, as a number of key retirements, including those of Professors Kinsloe, Breneman, and Bernhard, left glaring holes in the faculty ranks. (Another department head, Frederic Mavis, also departed that year, but to take a position at the Carnegie Institute of Technology.) Hammond found the job of securing qualified replacements to be a difficult and often impossible one. He frequently had to resort to shifting personnel from one assignment or department to another. This was especially true in the Department of Architecture, where most of the fine arts staff were pressed into service as instructors in engineering drafting courses.

The heaviest faculty attrition occurred at the very time undergraduate enrollment neared record levels, peaking in 1942 at over 1200. Total student enrollment was much higher. In addition to the three ESMDT courses, several large corporations contracted with the School to conduct six- and ten-month programs to train personnel for jobs as engineering technicians in the firms' defense plants. The two companies sending the largest groups of students, the Curtiss-Wright Corporation and the Hamilton Standard Division of the United Aircraft Corporation, recruited women—usually those with some collegiate education—for these programs, since most able-bodied men were either in the armed forces or already held civilian jobs that were vital to the war effort. One hundred four Curtiss-Wright "Cadettes" and an even greater number of Hamilton Standard students began their training at the College in February 1943. Two other aircraft manufacturers, the Glenn L. Martin Company and the Consolidated-Vultee Corporation, sent smaller contingents. Over 90 percent of these women completed their studies, a remarkably high proportion in view of the fact that the academic requirements of these short-term courses were no less rigorous than those of regular undergraduate courses. Many of these trainees went on to perform satisfactorily the same tasks for their employers that would have been delegated to baccalaureate mechanical, electrical, and aeronautical engineers in normal times.

Female engineering students were not exactly unheard of at Penn State before the arrival of the Cadettes and their counterparts from the other companies. Since the 1920s the Department of Architecture usually had one or two women in its architectural curriculum. In 1934 the Department of Electrical Engineering welcomed the first woman, Olga Smith of Philadelphia, to its ranks. Two years later, Ann Very, daughter of Penn State football hero "Dex" Very ('13, civil engineering) became the first coed to enroll in the Department of Industrial Engineering and in 1939 had the honor of being the first woman since Carrie McElwain Butts to earn an engineering degree from the College. No women completed requirements for degrees in electrical or mechanical engineering until the war years.

The School even boasted a woman on its faculty for a time. Russian-born Lidia Manson, who had studied mechanical engineering at the University of Paris, worked in the field of aircraft engines in France until Germany overran that country in 1940. She then fled to the United States, coming in 1941 to Penn State, where she worked with Paul Schweitzer as a research assistant in the diesel laboratory for two years. Manson became the first female to receive a graduate degree from the School of Engineering when she was awarded a Master of Science degree in mechanical engineering in 1943. While she never considered herself to be a pioneer for women in engineering, in a sense she was, and she encountered discrimination because of it. "I felt discrimination when the College would not let me continue studies toward a doctorate in engineering under the pretext that the curtailed war time curriculum was inadequate for such studies," Manson recalled later.[16] Undaunted, she left Penn State to pursue what would become a distinguished career in diesel engineering, gas turbine engineering, and environmental and energy systems development.

As the war intensified and the demand for engineers grew more acute, more women began choosing engineering as their course of study. The precise number of female undergraduates in the School of Engineering between 1943 and 1946 is difficult to determine. By mid-1943, President Hetzel confessed that he could no longer even give exact figures for total enrollment at the College, owing to the constant state of flux that surrounded academic affairs. A very conservative estimate of the number of women in the School would be about a dozen or so. The coeds often assumed an importance disproportionate to their numbers, however. In the fall of 1944, for example, five of the ten staff members of the *Penn State Engineer* were women.

In any case, undergraduates of either sex and the civilian trainees constituted less than half of the School of Engineering's total student load after June 1943. Just as in World War I, the Army and the Navy asked the School to provide technical training to enlisted personnel, mainly through the Navy's diesel and V12 programs and the Army's industrial management, electronics, and communications courses. By the time the war ended, the School had given instruction to a total of 2274 military trainees and 1077 civilian trainees, in addition to its regular student body. Had the number of baccalaureate students remained fairly constant throughout the war years, the School might not have so successfully discharged its responsibilities, given the steadily dwindling numbers of qualified faculty; but changes in selective service priorities enacted early in 1943 exposed most male engineering students to the draft. Enrollment plunged to about 330 by the end of the year, and barely 200 undergraduates were in residence by late 1944.

The School of Engineering's contributions to national defense were not restricted to the teaching field, although teaching certainly garnered most of the School's resources. "We have been able to make a less important contribution to the war effort through research in proportion to our potential capacity than we have in the field of instruction," Dean Hammond reported to President Hetzel in September 1943.[17] He attributed this imbalance mainly to the heavy classroom loads of the teaching staff, coupled with the scarcity of faculty and graduate student research personnel.

The Engineering Experiment Station had charge of most war-related engineering research. Personnel at the diesel laboratory did an extensive amount of consulting work for the Fairbanks-Morse Company, principal supplier of diesel power plants in submarines, and for the Navy itself during trials of diesel-powered surface vessels. The thermal laboratories in the early stages of war continued attempts to improve insulation for ships. These investigations were later expanded to encompass the development of better insulation materials for prefabricated military buildings that could be used in all types of climates. To assist in the building insulation project, the station in 1944 obtained $60,000 in federal assistance to construct a climatometer, or controlled weather room. Having a 30' × 24' floor and a 17' ceiling, the climatometer was large enough to contain a small house, whose insulative features could be studied under a variety of simulated climatic conditions, such as temperatures ranging from 130° to −60°F and rain, sleet, wind, and sand storms. Most of the research undertaken at the climatometer during the first few years of its existence concerned the testing of various dehumidifiers and drying procedures. The Navy and the United States Maritime Commission, looking to the day when huge fleets of war-surplus ships would have to be placed in storage, wanted to find the best methods of "mothballing" these vessels, that is, protecting them from the corrosive effects of moisture. The climatometer was also utilized in another important series of investigations aimed at discovering why so many mass-produced cargo vessels (Liberty ships) displayed an unpleasant tendency to split their hulls and sink without warning. Researchers eventually discovered that the steel used in the welded hull sections lacked sufficient manganese, causing the hulls to become extremely brittle when exposed to very cold temperatures. The Army Air Force sponsored yet another defense-related research project, this one involving the Department of Mechanical Engineering. At very high altitudes, where temperatures and oxygen levels are low, lubricants in older kinds of aircraft engines performed inefficiently. In the same methodical fashion that characterized their investigations of automotive engines in the 1930s, researchers quickly developed improved lubrication systems that were oblivious to the rigors of high-altitude flying.

A more indirect contribution to the nation's defenses occurred early in 1944 when the board of trustees authorized the formation of a Department of Aeronautical Engineering. Professor Peery headed the new department, the other faculty being Associate Professor Samuel K. Hoffman and Assistant Professor David J. Gildea. The department's offices and some classrooms were located in Unit E, while most of its equipment was in the Mechanical Engineering Laboratory. When the first six graduates received their Bachelor of Science degrees in aeronautical engineering at the June 1944 commencement, Penn State became the first and only institution in the Commonwealth offering such a course of instruction. Actually a fairly complete aeronautical engineering curriculum had existed since February 1942, at which time the trustees had approved aeronautical engineering as an official two-year option for junior mechanical engineering students. Instruction was provided in the areas of aircraft structures, aerodynamics, and aircraft engines. From the outset, Peery and his colleagues resolved to develop a program of superior quality. Their efforts first bore fruit in 1943, when Curtiss-Wright and the other defense contractors selected Penn State for their trainee programs in large part on the strength of the College's work in aeronautical engineering.

Well before the establishment of this new department, the job of administering the School of Engineering, with its myriad of defense courses and (until 1942) large number of degree candidates, taxed Dean Hammond's time and energy to their very limits. The affairs of the ESMDT National Advisory Committee also demanded much of his attention. President Hetzel came to Hammond's aid in 1942 by creating a new part-time administrative position, that of assistant dean of engineering. Appointed to this position was Royal M. Gerhardt, who was simultaneously named Professor of Architectural Engineering. A 1923 graduate of the University of Illinois, Gerhardt had served as an instructor in architectural engineering at Penn State until 1927, when he left to begin a fifteen-year career with the federal government, most of which he spent with the Public Buildings Administration. A modest, unassuming person, the new assistant dean brought to his post a ready smile and unfailing good cheer, qualities that were not conspicuous in his superior. Even without the pressure of war-time responsibilities, Dean Hammond was a stern taskmaster, whose reserved demeanor and humorless disposition were as notorious as his dedication to his profession.

By autumn of 1944, the pace of war work in the School of Engineering gradually began winding down. The civilian trainees had come and gone, as had the majority of military students. There was time now to think about what lay ahead for the School, about how to resolve the old problems of years past, and about how to deal with the new problems that the post-war years would inevitably bring.

5 New Directions: 1945–56

Harry Hammond contemplated the future of the School of Engineering with uneasiness. The national demand for engineers would surely increase once peace returned, yet the draft had depleted the student ranks of Penn State and other engineering institutions to the point where several years must pass before large graduating classes again became the norm. (The College awarded fewer than a hundred undergraduate engineering degrees between 1944 and 1946.) In contrast to the scarcity of seniors, Hammond expected massive numbers of students to enroll at lower levels. Many of these students would be former degree candidates returning to complete their studies, while others would be new students taking advantage of the "G.I. Bill of Rights," enacted into law by Congress in 1944, to obtain a college education at reduced costs. To accommodate the expansion of the student body, the School had to enlarge its physical plant, as well as acquire new and overhaul old laboratory equipment, most of which had suffered from overuse and deferred maintenance during the war. More faculty also had to be secured.

These problems were not unique to Penn State. Engineering educators throughout the country expressed uncertainty over the direction their profession should take after the war. In 1943 SPEE president Robert E. Doherty appointed a committee to assess the anticipated conditions of the post-war years and determine if changes in the methods and goals of engineering education were needed. Harry Hammond was named chairman of this panel. Summarizing the work of his committee in 1944, Hammond stated that "we find the general conclusions and recommendations of the Report of 1940 to be sound and applicable to engineering education under future conditions, as far as we can envisage them."[1] Hammond and his associates had spent most of their time making more specific suggestions as to how the objectives of the Report of 1940 could be met and formulating policies for dealing with the expected onslaught of returning veterans. Since these students would have many special requirements, the committee declared, the engineering schools' policies with regard to veterans "should be as liberal as possible without sacrificing sound educational standards or jeopardizing the interests of the individuals themselves." Hammond practiced at the local level what he preached nationally. Penn State's School of Engineering, as already noted, was the first institution to bring itself into substantial compliance with the guidelines accepted by SPEE in 1940, guidelines which were subsequently found to be viable for the foreseeable future, despite the disruptions caused by the war. In response to the second Hammond Report, the School of Engineer-

Western sector of the Penn State campus in the late 1940s, including all buildings of the School of Engineering. The structure in lower right, between College Avenue and the units, is the former College power plant, then housing the Petroleum Refining Laboratory and the diesel laboratory. (Penn State Collection)

ing and the Hetzel administration began mapping strategy to admit vastly increased numbers of students in the years immediately following the cessation of hostilities.

Nevertheless, Dean Hammond realized that, if his School was to preserve its standing as an important center for engineering education, it had to greatly expand the magnitude of its research program. Not only had engineering research played a vital role in the defense of the nation during World War II, but many defense-related developments—electronic communications and control systems, aerodynamics, new fuels and lubricants, and high tensile light alloys, to name just a few—had widespread peacetime applications. Educators, government authorities, and industrial leaders everywhere acknowledged research to be an increasingly important part of the academic phase of engineering. By 1946, forty-four land-grant institutions had established engineering experiment stations. However, these institutions could not maintain research at levels even remotely comparable to those of wartime without continued support of the federal government, which had sponsored the overwhelming share of research during the war.[2] Harry Hammond joined many of his colleagues from other land-grant schools in a campaign to convince Congress to enact legislation giving financial assistance to the engineering experiment stations in the same manner that it had for so long aided agricultural experiment stations.

Foundations for the Future

Pending the resolution of that campaign, how was Penn State to become a center for diversified engineering research, when in all likelihood the College would have to struggle just to meet the instructional needs of a soaring population of undergraduates? Dean Hammond was well on his way to answering that question even before Germany and Japan surrendered. Late in 1944 he persuaded Dr. Eric A. Walker, then assistant director of the Underwater Sound Laboratory at Harvard University, to come to Penn State when the war ended to become head of the Department of Electrical Engineering, a post vacant since Charles Kinsloe's retirement in July 1944. Meanwhile, Harvard was making plans to close the Underwater Sound Laboratory, which was operated in conjunction with the federal government's Office of Scientific Research and Development and the Department of the Navy. The Office of Naval Research deemed it unwise to abandon important defense-related scientific research and decided to carry on the activities of the USL at other locations. The Navy then asked Dr. Walker to put together a small staff and a laboratory of modest proportions and continue the work he had been doing on torpedoes. "I had in mind and the captain I was discussing it with in the Bureau of Ordnance had in mind about eight people and about $120,000 per year," Walker remembered. "Immediately I got in touch with Dean Hammond and asked, 'What about it?' He said, 'Fine, great, we'd just love to have it.' We examined a wing of the Electrical Engineering Building where there were about six rooms not being used and decided to put the lab in those rooms and build some shops. This was not a big deal, and we never even took it to President Hetzel."[3] No sooner had Hammond approved Walker's proposal to undertake torpedo research than the Navy and the Office of Scientific Research and Development concluded that a much larger facility was needed and suggested that the entire Ordnance Division of the USL move to Penn State. This time the matter had to be taken to Dr. Hetzel and the trustees, who gave it their unreserved blessing.

The principal tasks of the Ordnance Research Laboratory, as the facility was designated, were to develop and test various types of underwater weaponry, primarily torpedoes. In the spring of 1945 ground was broken for a brick laboratory building on a site near the intersection of North Atherton Street and the Bellefonte Central Railroad track. Construction progressed so rapidly that by September most of the former Ordnance Division's equipment and personnel were able to move from Harvard to the new State College location. The 150 or so scientific and technical specialists who accompanied the removal of the Ordnance Division to Penn State became faculty and staff members of the School of Engineering, since the ORL functioned as an administrative unit of the School. The Department of the Navy continued to exercise close supervision over the direction of the Laboratory's research, most of which was classified. Indeed, the Navy insisted on maintaining twenty-four-hour security, and armed guards accompanied some shipments of documents and hardware to and from Penn State. Adjunct facilities of the ORL consisted of a field testing station on a lake near Black Moshannon State Park (about 20 miles from the College) and field proving stations in Rhode Island and Florida.

The School of Engineering's acquisition of the Ordnance Research Laboratory was not paralleled by a similar expansion of instructional resources. In the fall of 1945, the School admitted its largest freshman class in nearly a decade. By September 1947, a total of over 2300 engineering students were enrolled. Sufficient space to accommodate this number did not exist on the main campus, so Penn State had made prior arrangements with seven of the fourteen state colleges to provide

The first phase of the Ordnance Research Laboratory nearing completion, 1945. (Penn State Collection)

living quarters and instruction for many freshmen of the engineering and other schools. The remaining freshmen, together with a portion of the sophomore class, attended classes at Penn State's four undergraduate centers: Altoona, DuBois, Hazleton, and Schuylkill (Pottsville). (The number of undergraduate centers was increased to seven in 1947–48.) It was a unique arrangement that permitted the School of Engineering to admit returning upperclassmen (most of whom were veterans), while ensuring that a steady supply of engineering graduates would be available four or more years hence.

It was also an awkward and expensive arrangement. Dean Hammond had already drawn up plans for more buildings, but his proposals were doomed by rising prices and the fiscal conservatism of Governor James H. Duff. A new building of substantial dimensions, designed to house the Department of Aeronautical Engineering and portions of the Departments of Industrial and Mechanical Engineering, formed the heart of Hammond's initial (1945) plans. When the College received only $3.6 million of a requested $8 million building appropriation in 1945, however, the dean reluctantly trimmed his request to the addition of wings for the Mechanical Engineering Laboratory and a fourth floor for the Main Engineering Building. Unfortunately, shortages of labor and materials prevented Penn State from beginning any new construction for nearly two years. By then, post-war inflation had so driven up costs that the original appropriation was no longer adequate, forcing President Hetzel to ask the Commonwealth for more funds. The General Assembly allocated $6 million, but Governor Duff, as part of a general campaign to reduce state spending, used his veto power to reduce that amount to a mere $750,000.

Before a final decision could be made as to how the money should be spent, President Hetzel died suddenly on October 3 from a massive cerebral hemorrhage. James Milholland, president of the board of trustees, assumed most of the duties of Hetzel's office until a permanent replacement could be found. One of Milholland's most unpleasant assignments was to bring the size of the new construction program within the limits of the appropriation. "We are now obliged to cut this absolutely basic building program still further," he told his fellow trustees, "at a time when requests for our educational services are at an all-time peak and still increasing."[4] Engineering was not the only school hampered by overcrowding. Conditions were even worse in other areas, and both the Mechanical Engineering Laboratory wings and a fourth floor for Main Engineering were deleted from the revised construction schedule.

Hammond fared better in his efforts to upgrade and enlarge his school's teaching staff. In a sense he continued where he left off before Pearl Harbor in his campaign to attract younger men to the faculty, which, he remarked in 1945, "is comprised of too large a proportion of older staff members."[5] By September 1946, he had been able to appoint several youthful new department heads along with numerous lower level personnel. Of the six department heads named between 1944 and 1946, only one, Milton Osborne, head of the Department of Architecture, was over forty years of age. In other ways, too, the composition of the teaching staff was changing. Hammond appointed individuals from a wide variety of educational backgrounds, noticeably lessening the academic inbreeding that had characterized the School of Engineering at Penn State and many other similarly isolated institutions in earlier years. Between 1936 and 1946, the engineering faculty grew from 102 to 136, yet the number of faculty members holding one or more degrees from Penn State declined by 2 (to 52) during that same period. Engineering faculty having doctorates rose from 2 in 1936 to 11 in 1946.

In yet another significant change of leadership that occurred early in the post-war period, Royal Gerhardt relinquished his position as assistant dean to become the College's new Assistant Registrar and Assistant Dean of Admissions. To fill the vacancy, Hammond in 1948 selected Professor of Electrical Engineering Earl B. Stavely. Stavely, who would prove himself to be an exceedingly capable administrator and a valued confidant of students, had received undergraduate and graduate degrees in electrical engineering from Penn State and had served continuously on its faculty since 1916.

Dean Hammond also found encouragement in the growth of research activities within the School. While the presence of the Ordnance Research Laboratory enabled Penn State to rank second among land-grant institutions in the dollar value of engineering research contracts during the late 1940s, the amount of money available to support research in the rest of the School was not large compared to the engineering research budgets of some other universities. Yet it did represent a step forward for an institution whose annual expenditures on engineering research had never surpassed $50,000 before World War II. In 1946–47, for example, the School's research budget totaled $1.8 million, of which the ORL received just over $1.4 million, mainly in the form of naval contracts. Most of the remaining $390,000, composed of assorted grants from the federal and state governments and private industry, was administered by the Engineering Experiment Station, whose full-time staff had grown from seven in 1940 to thirty-six in 1946. Some of the station's more extensive investigations were continuations or modifications of studies undertaken during the war. Researchers at the thermal laboratories renewed their efforts to develop improved insulation materials for use in prefabricated buildings for both military and civilian purposes. Civilian applications received top priority, as the post-war demand for high-quality housing that could be erected quickly and inexpensively reached unprecedented levels. The Department of Architecture also became involved in this field by developing a series of short courses for extension on the topics of home construction and buying and selling a house. In the diesel laboratory, Paul Schweitzer still led investigations aimed at increasing the combustion efficiency of diesel engines, having already attained significant breakthroughs in supercharging, scavenging, and weight reduction. Problems of lubrication in internal combustion engines of all types attracted the attention of members of the Departments of Aeronautical and Mechanical Engineering. Among new research projects undertaken by the Department of Civil Engineering were examinations of the potential of shale for use as a subgrade material in highway paving, and studies of treatment methods for industrial and sewerage pollution in streams and rivers. An important new project initiated by the Department of Electrical Engineering and the Experiment Station in cooperation with the Army's Watson Laboratories centered on studying what influence the ionosphere (the ionized layers of the upper atmosphere) had on high and low frequency radio transmissions. These investigations were directed by Professor of Electrical Engineering Arthur H. Waynick. Dr. Waynick was only one of many first-rate engineers and scientists who came to Penn State with the formation of the Ordnance Research Laboratory and later transferred to academic positions, where they had the opportunity to teach and carry on research on topics of personal interest. Aided by grants from federal civilian and military agencies, Waynick's atmospheric studies soon evolved into the Ionosphere Research Laboratory, headquartered in the Electrical Engineering Building. Its equipment included a transmitting station having a 4000-foot antenna supported on 400-foot masts and two receiving stations, one at the College farms and the other truck-mounted for easy mobility.

By far the largest addition to the School's research facilities (after the ORL) came in 1949 with the completion of the Garfield Thomas Water Tunnel. Situated west of the main entrance of the campus near the corner of North Atherton Street and Pollock Road, the tunnel was built to supplement the resources of the Ordnance Research Laboratory. It was named in honor of Lieutenant W. Garfield Thomas ('38, journalism), the first Penn State graduate to be killed in naval service during World War II. The largest circulating water tunnel in the world, it contained a 2000-horsepower motor driving water through the test section at speeds up to 60 feet per second. The Navy's Bureau of

Garfield Thomas Water Tunnel under construction. (Penn State Collection)

Ordnance provided the $2.15 million needed to construct the tunnel, which was used in testing underwater shapes, such as torpedoes, submarine hulls, and propeller designs. The addition of the water tunnel brought even more naval contracts. By the time the tunnel reached its full operating potential in 1950, the Ordnance Research Laboratory had a yearly research budget of over $2 million and employed some 230 full-time personnel. In speaking at the dedication of the water tunnel, held October 7, 1949, Assistant Secretary of the Navy John T. Koehler explained the Navy's rationale for investing so much money in research facilities at Penn State. "The last war proved conclusively that it is not possible to conduct basic research during hostilities and to convert knowledge gained thereby into weapons soon enough to have a decisive effect," Koehler affirmed. "Investment of military dollars in scientific research in peacetime pays dividends in the form of time saved in wartime. And saving time during a war means saving lives, material, and money."[6]

A direct response to the heightened level of research activity in the School of Engineering was a dramatic increase in the graduate student population. During the mid-1940s, one of the main topics of discussion in engineering education circles pertained to the merits of a five-year baccalaureate curriculum. By 1946, Cornell, Columbia, Dartmouth, and a half-dozen other universities had taken steps to adopt the longer course of study. Citing the recommendations of the Wickenden and his own reports, Harry Hammond steadfastly opposed the change, at Penn State and elsewhere. He reasoned that a five-year curriculum would only prompt students to specialize at the very time when they ought to be educated in the broad principles and fundamentals of the profession. A sounder plan according to Hammond would be to provide a four-year curriculum for all students but give increased emphasis to graduate work, where specialized studies were more appropriate. He estimated that 30 to 40 percent of all seniors should undertake advanced studies immediately after graduation, and he set this as the goal of his school. Only about 5 percent of Penn State's engineering undergraduates during the immediate pre-war years had elected to go on to do graduate work, a figure that was in line with the national average. Dean Hammond did concede that a five-year curriculum in architecture and architectural engineering was necessary, since the National Architectural Accrediting Board refused to accredit any four-year curriculums after 1947. Penn State was then one of only three or four schools still on a four-year schedule. At Hammond's recommendation, the College trustees approved the change, effective with the freshman class entering in September 1948.

With the exception of the Department of Architecture, the School of Engineering retained the four-year curriculum. In time, most of the institutions that had experimented with longer programs returned to the traditional four-year curriculums, too. In virtually all engineering schools, graduate education was growing. At Penn State, 116 graduate students were in residence in 1950–51, triple the number of 1946–47. The Experiment Station no longer exercised administrative supervision over graduate students, who instead were attached to their respective departments. All departments offered courses of study leading to the Master of Science degree, and most had doctoral programs. Never before had financial assistance for graduate work been so readily available. Supplementing departmental teaching assistantships and Graduate School Fellowships was a growing number of fellowships funded by General Motors, DuPont, and other large firms.

Overcrowded conditions plagued graduates and undergraduates not only in the classrooms during the late 1940s. With total undergraduate enrollment at Penn State's main campus ballooning from 5950 in 1940 to over 9500 by 1949, a critical need arose for more residential housing. The community of State College, heretofore little different from numerous other small, central Pennsylvania towns, became the site of a frenzied effort to build low-cost private housing. While new homes were being built in the borough, the College, lacking funds for new dormitories, had to make do with military housing left over from the war. The cluster of barracks (of which Nittany Halls is the last remnant) on the east side of the campus and the tiny mobile homes scattered over "Wind Crest" (much later the site of Pollock and South Halls dormitory complexes) may have left a great deal to be desired as comfortable dwellings, but at least they did provide necessary shelter. Somehow, most students—many of whom had wives and children to support—made it through the crisis.

With the exception of discontent caused by overcrowding on and off the campus, the most commonly voiced complaint among engineering students concerned the quality of instruction they were receiving. Criticism was directed mainly toward the lower faculty ranks, populated as they were by many new and inexperienced instructors who had been hired to meet the ever larger enrollments. Strong sentiment existed among the student body for the creation of a short course in educational methods, wherein new teachers could learn how to transmit their subject material more effectively and better stimulate their classes. "Before a graduate engineer starts to work, he is put through a training period to be familiarized with the functions of the employing organization," stated a *Penn State Engineer* editorial in April 1949.

"By the same token, new instructors should be taught the functions of teaching so that they can do a better job." Apparently the dean and faculty believed more urgent matters demanded their attention, and the issue remained a source of sporadic student dissent.

These and other, more minor dissatisfactions notwithstanding, student morale in the Engineering School was as remarkably high as the failure rate was low. In his annual reports for these years, Dean Hammond never tired of pointing out that, as far as he could determine, the drop-out rate among engineering students had never been lower. He was especially proud of the former Curtiss-Wright and Hamilton Standard female trainees who returned to the College. Having had an opportunity to gain practical experience during the war, they came back to complete the requirements for their degrees and become fully qualified engineers. Only one aspect of the academic performance of the students of his School disturbed Hammond, and that was the difficulty many of them encountered in taking the English Proficiency Test. Beginning in 1946, in compliance with a nationwide trend toward strengthening technical students' communications skills, the School of Engineering required all juniors to pass this test as a prerequisite for graduation. Those who received a failing grade had to take a remedial course in English during their senior year. Why 15 to 20 percent of the junior class regularly failed the proficiency examination and more than a few students encountered trouble with the remedial course mystified Dean Hammond, who never was able to find a satisfactory solution to the problem.

A different kind of student—one no less determined than the degree candidates—enrolled in the extension classes. To be sure, engineering extension still had charge of instruction at the undergraduate centers, but most students attending those centers were seeking baccalaureate degrees and eventually would go on to the main campus. The majority of students who registered for engineering extension classes attended one of the ten "technical institutes," the new name for the evening branch centers, located throughout the state. The institutes now offered instruction during the day as well as at night and in a wider selection of subjects. As in the past, no college credit was attached to these courses, whose curriculums were still oriented toward practical applications in local industries. Although they were not offered at all locations, the most popular institute courses were "Industrial Electricity," "Mechanical and Production Tool Design," "Building Construction," and "Production Management." In addition, engineering extension provided its usual array of short courses and special institutes for firms and trade associations doing business within the

Commonwealth. In 1947, for example, when the Pennsylvania Railroad decided to begin replacing its huge fleet of steam locomotives with diesel-electric units, it arranged to have the extension department conduct classes in diesel locomotive maintenance for hundreds of its shop employees. The undergraduate centers, the informal courses, and the correspondence division consistently enrolled over 3000 students yearly in the late 1940s and early 1950s. In 1949 Kenneth L. Holderman succeeded Edward L. Keller as director of engineering extension. Holderman, a 1931 graduate of Penn State's architectural engineering curriculum, had been a member of the extension faculty since 1941. Keller moved up to become executive assistant to J. Orvis Keller, who still headed the College's Central Extension office.

Also in 1949, for the first time in its history, the Department of Engineering Extension offered a course for graduate credit. Thirty-four bridge and structural engineers took a civil engineering course, "Statically Indeterminate Structures," at Harrisburg in connection with their work in building the eastern segment of the Pennsylvania Turnpike. Dean Hammond encouraged the further development of graduate-level extension courses, but to his chagrin, he discovered that he was unable to modify most of the restrictions that the Graduate School had placed on granting credit for off-campus studies. Pending the relaxation of these stipulations, graduate work in engineering extension held little hope of expansion, despite the growing demand for such courses by engineers throughout the Commonwealth.

The achievements in research and extension in no way mitigated the School's desperate need for more instructional space. As late as 1948–49, over a thousand freshmen engineering students still had to be "farmed out" to the state colleges for lack of room on Penn State's main campus. "We are not able to render the service to the Commonwealth and to the nation that we ought to be rendering," Dean Hammond lamented in his annual report in 1949. In describing conditions in the School, he noted that

> many faculty members are crowded together in offices where it is impossible to confer properly with students or to carry on other duties. Some rooms have as many as 6–9 faculty members. . . . Both of the libraries—Architecture and Engineering—are over-flowing into instructors' offices. There is no room for students to study between classes, and they are seen studying on stairs or propped up against corridor walls. The Dean's office cannot accommodate any more clerical workers for lack of space, thus all occupants of the office are overworked. The fact is that a physical plant adequate for about 1200 engineering students is now serving an on-campus enrollment of more than 1800.

The situation with regard to laboratory equipment was nearly as bad. Not only was there an insufficient quantity of equipment for the students; much of the hardware that the School did possess was woefully obsolete. Penn State's School of Engineering spent approximately $5000 annually on new and reconditioned laboratory gear during the late 1940s, or only about one-third the amount expended by the engineering schools of other large land-grant institutions. "We are using for instruction machines and instruments which are out of date and excite the wonder of visitors from better equipped schools," Hammond observed.[7]

Some temporary relief was finally obtained in 1950, soon after the Commonwealth reestablished the General State Authority, defunct since the war. Acting President Milholland promptly submitted a $15 million list of buildings that the College hoped to erect over the next three years. The GSA approved a building schedule totaling only $8.9 million, but even so, the amount represented the largest building appropriation in Penn State's history. Included in it were $1.16 million for additions to the Mechanical Engineering Laboratory and $157,000 for the construction of a fourth floor for the Main Engineering Building. Work began in August 1950, with Henry E. Baton, Inc., of Philadelphia, as general contractor. The enlargement of the Mechanical Engineering Laboratory was completed late in 1951. Designed by the Pittsburgh architectural firm of Altenhof and Brown, two new three-story wings contained space for offices, computation rooms, and a large lecture room. Laboratories for air conditioning and refrigeration, lubrication, and carburetion were also housed in the new wings. To make way for a larger parking area behind the Laboratory, workers razed the 45-year-old Engineering Unit F. Few tears were shed over the old frame structure's demolition, for it had been one of the campus's most prominent eyesores (not to mention a dangerous firetrap) ever since its construction in 1907 as a "temporary" solution to the School's need for more space. The addition of a fourth floor to the Main Engineering Building eliminated what a generation of architectural students had derisively labeled "the Black Hole of Calcutta," a name for the third-floor drawing rooms that stemmed from almost intolerable lighting that prevailed there. These rooms were now moved to the new fourth floor, where skylights provided generous amounts of natural illumination. Further expansion of the School's physical plant came in 1951 when the Navy supplied funds for the construction of a second floor over the entire Ordnance Research Laboratory and a third floor over part of the building. The beginning of the School's long overdue construction program marked the last major accomplishment of the Hammond administration. At the close of the 1950–51 academic year, the 67-year-old dean retired.

Harry Hammond had brought some fundamental changes to the School during his tenure as dean: upgrading and modernization of the curriculums, an enhanced role for engineering research, and an infusion of young, innovative faculty members, to name just a few. Even more than Robert Sackett, Hammond was a giant in national engineering education affairs, co-authoring two national reports that influenced the direction of technical education for many years afterward. In 1945 the Society for the Promotion of Engineering Education honored Hammond with its prestigious Lamme Award. Eight years later he won an even rarer distinction by being elevated to honorary life membership in the Society, which had just changed its name to the American Society for Engineering Education. Hammond also participated in the movement to have Congress establish an agency to support scientific and engineering research. The land-grant schools had worked to secure an engineering counterpart to the Hatch Act for several decades but had been unable to exert sufficient political pressure to achieve their goal. Shortly after World War II, they affiliated with other institutions to form the Engineering College Research Association, which in turn joined with the scientific community in an intensive lobbying campaign, climaxing in 1950 when Congress enacted legislation creating the National Science Foundation. As the author of two landmark studies of engineering education, both of which called for a greater emphasis upon research, Hammond contributed significantly through public testimony and more covert manueverings to the ultimate passage of the NSF bill. A fitting climax to Hammond's Penn State career occurred in May 1951, when the College presented its first annual Distinguished Alumni awards—the highest honor that Penn State could bestow upon its graduates. Three of the five individuals receiving the awards were alumni of the School of Engineering: Charles E. Denney, Bayard D. Kunkle, and Clarence G. Stoll.

After leaving Penn State, Hammond continued to make his home in State College. Unfortunately, he was denied the opportunity of enjoying an extended retirement, succumbing on October 21, 1953, to a lingering illness. Perhaps his most appropriate epitaph came from one of his former students, who said simply, "The name Harry Hammond is another term for engineering education."[8]

Eric Walker and an Era of Modernization

Appointed as the new dean of the School of Engineering was Eric Walker, Harry Hammond's personal choice for the job. Hammond had informed Walker several years earlier that he hoped the young electrical engineer would eventually succeed him as dean, and the two men had subsequently reached an informal understanding regarding the matter. Born in England, Eric Walker had spent much of his youth in Wrightsville, Pennsylvania. In 1928, he enrolled as an engineering student at Harvard, which he chose over Penn State only because the Ivy League school's scholarship offer was more attractive. After receiving his bachelor's degree, Walker continued his studies at Harvard, first earning a master's degree in business administration and then a doctorate in science. After teaching in the electrical engineering departments of Tufts College and the University of Connecticut, he returned to his alma mater in 1942 to participate in the underwater sound research program. Walker assumed his new duties as dean in September 1951, at which time Arthur Waynick became head of the Department of Electrical Engineering. Professor Waynick continued to supervise activities at the Ionosphere Research Laboratory as well. Dr. Gilford G. Quarles was named head of the Ordnance Research Laboratory, where he had previously been associate director. Both Waynick and Quarles had already been occupying their new posts on an acting basis for the past year, while Walker was on leave as executive director of the Defense Department's Research and Development Board.

Walker brought a refreshing change in personal style to the dean's office. For example, a colleague of both Hammond and Walker recalled that "Dean Hammond's faculty meetings were pretty staid. They were pretty solemn affairs." In contrast, "at the first meeting Eric Walker had, the gang assembled, and typically most of them sat in the back of the room, and about the first six rows were all vacant up front. Dr. Walker went up onto the platform and with a very straight face said the first announcement he wanted to make was that only those sitting in the first five rows of the auditorium would be eligible for salary increases next year."[9] This lighter touch, while perhaps not so visible to students, was much appreciated by the teaching staff and fostered a spirit of camaraderie that had been lost during the upheavals of post-war expansion.

Just as the School of Engineering underwent a change of leadership, the College's new chief executive was acclimating himself to the routine of Old Main. The trustees had not chosen a successor to Ralph Hetzel until 1950, when they finally selected Dr. Milton S. Eisenhower. Eisenhower, 51-year-old brother of General (and soon to be President) Dwight D. Eisenhower, came to Penn State from Kansas State University, where he had also been president. He brought with him unbounded enthusiasm and a considerable store of new ideas for

Eric A. Walker (Penn State Collection)

enlarging the College's role as one of the nation's major centers of higher learning. Some of these ideas were to have a direct bearing on the future of the Engineering School.

Dean Walker's administration commenced just after the post-war flood of engineering had crested. In 1950–51, the School permitted freshmen to begin their studies on the main campus for the first time since 1945. Many students began to fear that the large classes that had preceded them in the late 1940s would glut the job market for engineers; but as the decade of the late 1950s wore on, their worries proved groundless. The onset of the Korean War in 1950 caused many government and industrial recruiters to step up their campaigns to attract young engineers. Non-defense-related areas of the economy remained strong, too, so few graduating seniors experienced trouble in obtaining satisfactory employment.

The Korean War did adversely affect some activities in engineering extension, as the threat of the draft and the lure of high pay in defense industry jobs cut into enrollments at the technical institutes. Enrollment at the undergraduate centers, by contrast, held steady at about 300 students per year, or about 25 percent of the total student body at the centers. The unsettled situation in the Graduate School, where Dean Harold K. Schilling remained unconvinced of the wisdom of easing the restrictions on the awarding of credit for non-resident instruction, still frustrated efforts to offer more graduate courses. Extension nonetheless continued to conduct advanced courses in bridge design at Harrisburg and even added new courses in electronics and vibrations at Erie.

Walker, like Hammond, recognized the value of engineering research as a means of expanding man's sphere of useful knowledge. Under his leadership, research activity at the Ordnance Research Laboratory expanded until by 1953–54 the ORL's research budget amounted to nearly $2 million, or approximately two-thirds of the funds spent on engineering research at Penn State that year. Acoustics, electronic circuitry, and fundamental hydrodynamics still dominated the interest of ORL researchers and their Navy patrons.

Research expenditures in the rest of the School climbed steadily, too. For example, the Ionosphere Research Laboratory, separated from the Engineering Experiment Station in 1952, saw its annual budget rise from about $100,000 that year to nearly $1 million by the end of the decade. The National Science Foundation and other civilian and military agencies provided the bulk of this support. In the early 1950s, the IRL shifted the emphasis of its investigations to the study of the chemistry and physics of the upper atmosphere, a field previously little explored by scientists and engineers. Penn State soon became the leading international center for such studies, as dozens of distinguished researchers from Great Britain, France, Australia, India, and other countries came to the IRL for advanced work. Because the Laboratory was a scientific as well as an engineering facility, the School of Engineering for the first time was able to offer opportunities for extensive interdisciplinary research and attracted faculty and graduate students from physics, chemistry, mathematics, meteorology, and other departments. The Laboratory became the chief source of support for thesis research projects in electrical engineering. The discoveries made by IRL researchers led to a fundamental theoretical understanding of the chemistry and physics of the ionosphere and had important applications in solving problems related to space travel, man-induced disturbances of the upper atmosphere (particularly the ozone layer), long-wave radio transmission, and in other fields.

After the IRL was detached, the Experiment Station experienced a reorganization. Its name was changed to the Department of Engineering Research in order to more clearly convey the kind of research and development work that was carried on there. The term "experiment station" connoted testing activities, when in reality testing of materials had taken on secondary importance in favor of more scientific studies, most of which were still related to heat transfer in building materials and internal combustion engineering. In conjunction with the adoption of the new name, Elmer R. Queer, acting director of the Experiment Station since 1950, was appointed head of the Department of Engineering Research. Queer was a veteran Penn Stater, having received both bachelor's and master's degrees in electrical engineering from the College in the 1920s and having served on the faculty at the Experiment Station since 1928. An expert in insulation and ventilation, it was Queer who had directed the studies after the war on the best methods of mothballing the Navy's surplus vessels.

Research projects undertaken by the individual departments were still rather limited in number and scope, however. As in past years, the responsibilities of undergraduate instruction placed severe restrictions on faculty research opportunities. "We are operating at an emergency load with a staff shortage," Norman R. Sparks, head of the Department of Mechanical Engineering, stated in a typical complaint to Dean Walker in 1954. "Work for extension, work for correspondence courses, for graduate research supervision, all of which are legitimate functions, are habitually carried as an overload."[10] Staff shortages and heavy teaching loads were not the only obstacles to increasing research. Walker discovered that many of the senior faculty had a tacit understanding that research should be left to members of the Experiment Station and that the instructional staff should content themselves exclusively with teaching, except for an occasional investigative foray that was more akin to tinkering than to serious research. This attitude was perhaps more understandable during an era of slim budgets and classrooms overcrowded with undergraduates—conditions that still persisted, of course, but not to the extent of earlier years. Furthermore, the older members of the teaching staff had never faced the "publish or perish" syndrome that was now becoming an increasingly prominent fact of academic life. Dean Walker often had difficulty in convincing his colleagues that research comprised a legitimate and worthwhile part of their activities. He cited a typical case of an engineering professor who was "a perfectly competent fellow, a good teacher, but to get him to do any research, I had to hold his hand all the way up through buying the equipment, setting it up, taking measurements."[11] After the research was concluded, the dean had to go to extraordinary lengths to see that the results were published, since his colleague lacked sufficient confidence in his own work to disseminate these findings himself.

Walker believed that the only way Penn State could attract engineering teachers of consistently high quality was to promise them ample time to pursue research. The salaries accorded faculty members still could not compete with those of government, industry, and some other engineering schools, but an opportunity to do research work in a field of personal interest might be something no other employer could offer. Harry Hammond held a similar view, but the war had interfered with his efforts to encourage research. Moreover, Hammond lacked the drive and personal magnetism of Eric Walker, an ambitious man who knew precisely what he wanted and would toil ceaselessly until he achieved it, even if he occasionally had to resort to extraordinary means.

A case in point involved the nuclear reactor, easily the most significant addition made to engineering research facilities during Dr. Walker's tenure as dean. With the end of World War II had arrived the dawn of the atomic age. The role of the atom would become so pervasive, Harry Hammond had predicted before his retirement, that engineers not acquainted with the fundamentals of nuclear energy would before long find themselves in the dark ages of science and engineering. Beginning in 1952–53, the School of Engineering required every undergraduate to take an elementary course in nuclear energy as part of the general sequence in physics. The following year the Department of Engineering Mechanics began offering a pair of three-credit electives for juniors and seniors in the foundations of nuclear engineering. Eric Walker foresaw the day when the School would have to teach nuclear engineering as a curriculum in its own right. He started giving some thought to the subject matter and approaches and discussed them with his friend Dr. William M. Breazeale, a physicist at the Oak Ridge National Laboratory and a former Professor of Electrical Engineering at the University of Virginia. Breazeale suggested that Penn State should have its own reactor, which he offered to design (based upon a swimming-pool-type reactor he had already designed at Oak Ridge) and which he contended could be built for less than $250,000. Walker was skeptical at first, since the federal government through the Atomic Energy Commission (AEC) controlled all fissionable materials and exercised close supervision over the construction and operation of all reactors. Even assuming Penn State could build a reactor for the relatively modest sum Breazeale quoted, could it get government approval to go ahead with the project in the first place? On the other hand, the dean had no doubts about the tremendous value a reactor would have as a research and teaching instrument for the entire College, not just the School of Engineering. He took the proposal for the reactor to Milton Eisenhower, who liked the idea immediately. It fit in perfectly

President Dwight Eisenhower at the nuclear reactor, 1955. Also present are, from left, William Breazeale, Milton Eisenhower, and Eric Walker. (Penn State Collection)

with the President's desire to advance Penn State to the front rank of the nation's institutions of higher learning. After obtaining trustee consent, Eisenhower and Walker presented their plans to the Atomic Energy Commission, which was preparing a campaign to broaden the peaceful applications of nuclear energy and had already approved the construction of a very small (less than ten-kilowatt) research reactor at the North Carolina State College. The Commission believed that the nation's colleges and universities should assume a greater share of the responsibility for the training of nuclear scientists and engineers in anticipation of more widespread non-military utilization of nuclear power. (Until this time, the AEC itself supervised the training programs for nuclear specialists.) Therefore, it expressed a willingness to cooperate with Penn State in building the reactor, if the College would pay all costs of construction and operation, with the exception of the fissionable material, which the Commission would supply. The board of trustees agreed to this condition, and work on the reactor facility got underway in May 1953.

The reactor, which was to have an initial maximum output of 100 kilowatts, was located in what was then an isolated area on the east side of the campus. In order to keep costs to a minimum, the engineering faculty and staff were called upon to contribute to the design and construction of the facility. Professor of Architectural Engineering Louis Richardson prepared blueprints for the building housing the reactor, and Professor of Civil Engineering Gerald K. Gillan designed the reactor pool. Shop personnel in the Ordnance Research Laboratory fabricated many of the mechanical parts, with Clarence Ellsworth and Clyde Witman of the Department of Industrial Engineering crafting some of the more exotic components. Rupert Kountz of the Department of Civil Engineering helped resolve many of the knotty problems surrounding waste water and handling. Supervising the whole affair was William Breazeale, who had come from Oak Ridge at Dean Walker's invitation and on July 1, 1954, became Penn State's first Professor of Nuclear Engineering.

Work on the reactor facility proceeded on schedule. On February 22, 1955, dedication ceremonies were held, with Admiral Lewis L. Strauss, chairman of the AEC and President Dwight Eisenhower's personal advisor on atomic energy, delivering the main address. The subject of nuclear research at Penn State attracted even more public attention in June, when the President himself, at the urging of his brother Milton, came to address the graduating class and personally inspect the reactor. During the ceremonies in the reactor building, all eyes fixed on "Ike" as he stood with Dr. Walker and Milton Eisenhower at the reactor's control panel. "Where is the button I push to start it?" the President asked Walker. "Mr. President, the button you push is in Washington," replied the dean. "We don't have the fuel," he explained to an incredulous Eisenhower, pointing out that officials at the Atomic Energy Commission "down there in Washington have the fuel on a truck and they can't ship it, because they don't know how to do it."[12] The President pushed the button anyway, and upon his return to the nation's capital immediately looked into the matter of the missing fuel. The fuel elements arrived at Penn State the following week.

The problem of transporting fissionable material was only one indication of the pioneering nature of Penn State's attempt to obtain a nuclear reactor. Another was the difficulty encountered in acquiring a licensed operator for the reactor. The Commission had never before granted an operator's license to any individual, since it had heretofore enjoyed total control over the operation of all the nation's reactors. The only other reactor then in the possession of an institution of higher learning, the one at North Carolina State College, had been completed in 1953 under a "study agreement" with the AEC. In 1954, however, Congress revised the Atomic Energy Act, easing restrictions on the building of reactors in order to encourage the private sector to take a more active part in the development of nuclear energy. This legislation required the AEC to license both reactors and operators. The dilemma finally resolved itself when Dr. Breazeale, with the Commission's tacit approval, wrote an examination for an operator's license and then gave it to himself. He passed the test easily, and on August 15, 1955, the reactor at last reached criticality, the point at which a self-sustaining reaction occurred. In this curious way, Breazeale became the first operator licensed by the AEC, and for the next several years Penn State, as the first American college or university to possess a licensed reactor, became a center for the training of nuclear reactor personnel. The North Carolina State reactor, which had first reached criticality on September 5, 1953, was not formally licensed until Penn State's facility had been in operation for several weeks. (Penn State nevertheless received license R-2 rather than R-1.)[13]

Penn State also played a vital role in indoctrinating scientists and engineers from other countries in the wonders of the atom. In his June 1955 commencement address, Dwight Eisenhower used the reactor as the springboard for one of his historic "Atoms for Peace" messages. In that speech, he called for greater efforts to harness nuclear fission for peaceful purposes and pledged to put the research reactors of the United States at the disposal of qualified technical personnel from all free nations. Penn State pledged its full cooperation and in April 1956 joined with the Atomic Energy Commission, Argonne National Laboratory, and North Carolina State College to form the International School of Nuclear Science and Engineering. Under this program, scientists and engineers from other countries spent a semester in resident training at Penn State or North Carolina State before going on to Argonne for additional study. By January 1959, when a decline in enrollment led the School to consolidate its program at North Carolina State, 179 students representing 39 nations had participated in the ISNSE using the Penn State reactor.

The School of Engineering initially utilized the reactor for research and the training of graduate students. While the facility was available to all qualified faculty members, it was to remain closely identified with engineering applications and, unlike reactors at several other institutions, never became the centerpiece for a broad program of fundamental research in nuclear physics. Nor did it lead to the immediate establishment of an undergraduate curriculum in nuclear engineering. Writing in the *Penn State Engineer* in October 1955, Eric Walker declared that "the men who design and build reactors for the production of energy or the radiation of specimens or the production of isotopes are confronted with the same old problems in mechanical design, heat transfer, metallurgy, and materials that engineers have been confronted with for many years." Walker shared the conservatism of most engineering educators of that era with regard to the new field of nuclear engineering and urged students first to acquire a solid background in, say, mechanical engineering and then concentrate on nuclear physics and related topics in graduate school. Virtually all of America's nuclear engineers had followed that route. Furthermore, even if Penn State did wish to award bachelor's degrees in nuclear engineering, Dr. Walker claimed that the field was so new that he would be unable to secure a sufficient number of qualified instructors at the salaries normally attached to academic positions. In the end, the decision was made to forego initiating a baccalaureate curriculum but to assess the advisability of introducing a graduate program.

The construction of a nuclear reactor represented yet another milestone in the history of engineering education at Penn State. But by the time the reactor went into operation, the institution was no longer calling itself The Pennsylvania State College. Milton Eisenhower believed that The Pennsylvania State University was a more fitting title for an institution that engaged in a broad range of teaching, research, and service activities. By 1953 he had won over the trustees to this point of view, and in November of that year the Centre County court gave its official consent to the name change. Shortly thereafter the trustees approved the substitution of "college" for each of Penn State's undergraduate schools, with the exception of the School of Engineering, which upon Dr. Walker's recommendation became the College of Engineering and Architecture.

Curricular and Administrative Changes

Eric Walker did not confine his attention to upgrading engineering research. Several curricular innovations, all of which became important assets to the College of Engineering, also had their genesis during the Walker years. One of the earliest innovations was the cooperative degree program, established in September 1952 by the School of Engineering and several small liberal arts colleges. This five-year course of study was aimed at those students who were uncertain of their interest in engineering, as well as those who were sure they wanted to become engineers but desired a more broadly based education. Participants spent their first three years at a liberal arts school pursuing general studies that included moderate amounts of science and mathematics. At the end of their junior year, the students transferred to Penn State, where, upon completion of two years of engineering studies, they received bachelor's degrees in both liberal arts and engineering. The liberal arts institutions benefited by increasing their enrollments and reducing their need to maintain laboratory and other technical facilities having high fixed costs. Having some lower classmen study off-campus made more room available at University Park for those students who wished to enroll in one of the regular four-year engineering programs. Four colleges initially joined Penn State in this program: Albright, Gettysburg, Lycoming, and Westminster. By 1959, Elizabethtown, Muskingkum (Ohio), St. Francis, and St. Vincent colleges had been added to the group. The number of participating schools varied in subsequent years, but by 1980 over 180 persons had received undergraduate degrees through this program, and a total of eighteen liberal arts and state colleges were participating in it. In 1960 the Colleges of Engineering and Liberal Arts entered into a similar arrangement, known as the dual degree program, whereby a student enrolled in the University's College of Liberal Arts for the first nine terms (three academic years) and took coursework in the humanities, social sciences, mathematics, physical sciences, engineering graphics, and engineering mechanics. The student then completed six terms of professional and technical studies in his major in the College of Engineering, after which he received both B.A. and B.S. degrees. More than fifty students earned dual degrees in the first two decades of this program's existence.

In September 1953, the School of Engineering introduced another new course of studies, this one oriented toward students having a very different set of interests and goals from those taking combined liberal arts and engineering studies. Dean Walker had long maintained that too many engineers "are not doing work worthy of the profession." By that he meant too many individuals holding bachelor's and even advanced degrees "are carrying out routine tests, calculations, and drafting which technicians could do as well if not better." In Walker's opinion, engineering was a "creative science," yet too many engineers filled essentially non-creative posts owing to a shortage of qualified personnel.[14] Walker and Director of Engineering Extension Kenneth Holderman believed that Penn State's technical institutes were no longer sufficiently comprehensive in some fields and were losing many potential students to undergraduate programs. Walker, Holderman, and their colleagues began working on a plan to upgrade several of the one-year full-time and three-year part-time technical institute curriculums into two-year full-time courses of study having a continued emphasis on practical rather than theoretical subject matter. The Hammond Report of 1944 had called for baccalaureate degree-granting institutions to become involved in training engineering technicians, and before the end of the war the Engineers' Council for Professional Development had set up formal accrediting procedures for what it officially classified as "programs of the technical institute type." Nevertheless, most schools having undergraduate curriculums in engineering expressed reluctance to adopt purely technical curriculums of one or two years' duration. As late as 1952, fewer than a half-dozen of these schools offered accredited instruction at that level. Like Penn State, most already had a strong tradition of sub-baccalaureate technical education. New engineering technician programs were most numerous at community and junior colleges and at public and private technical institutes. The states of New York, California, Florida, and Texas were especially active in this area. The kind of program envisioned for Penn State resembled the two-year technical institute curriculum introduced at Purdue University in 1942 under the supervision of Charles W. Beese. However, in order to highlight the distinction between the older technical institutes and the new two-year "engineering technology" curriculums, Holderman proposed that the new curriculums lead to Associate in Engineering degrees. Only one baccalaureate degree-granting institution—the University of Houston—was then conferring associate degrees (Associate in Science) on graduates of its technical institute-type programs, and even among community colleges and technical institutes, the Associate in Engineering degree was a novelty. Most institutions simply issued certificates of completion.[15] Holderman believed the associate degree was essential if the engineering technology courses were to compete successfully with undergraduate studies. Even after the University's board of trustees approved the use of this degree, other obstacles had to be overcome. Dr. Walker recalled later that "our major job was getting students, principally because the high school counselors knew nothing about engineering. So Ken Holderman and I held dinners for eighty percent of the high school counselors in the state in groups of one hundred, gave them a talk about these curriculums, where they would be offered, etc., and eventually we built up a student body."[16] Walker and Holderman were very persuasive. In the fall of 1953, over four hundred students enrolled in the first two associate degree courses, "Drafting and Design Technology" and "Electrical Technology," at ten locations throughout the Commonwealth. No associate degree classes were held at University Park, as the main campus came to be designated during the Eisenhower era, although summer sessions were introduced there later. The growing demand for engineering technicians, the abundant experience with technical institutes, and ASEE and ECPD recommendations all came together to provide the initial impetus for the development of Penn State's engineering technology program. Further stimulus came in 1957 when the Pennsylvania legislature rejected Governor George Leader's scheme to create a system of state-supported community colleges that were to offer both liberal and technical studies. In the absence of these two-year schools, the University accelerated the expansion of its branch campuses as well as the engineering technology and other associate degree programs. By 1960, the College of Engineering was awarding over 1500 Associate in Engineering degrees annually. Thus some of the College's prominence in this field can be said to have been gained by default, as Penn State was left little choice but to expand to meet the educational needs of the Commonwealth. On the other hand, besides being shaped by state and national events, the College of Engineering's associate degree program served in some ways as a model for other baccalaureate institutions that were just entering the field of technician training. Throughout the late 1940s and 1950s, Professor Holderman was a nationally recognized leader in the ASEE's Technical Institute Division and played a particularly influential role in reorganizing accrediting criteria. When in 1958–59 the ASEE moved to achieve uniformity for accrediting purposes in education at the technical institute level, it was no coincidence that most of its recommendations—that the length of the curriculum be two years, that the word "technology" be in course titles, and that an Associate in Engineering be the usual degree granted, for example—had been in effect at Penn State since 1953. For his work in the Technical Institute Division (since 1971 the Engineering Technology Division) and for his leadership in piloting the College of Engineering's engineering technology program, Holderman received in 1960 the James H. McGraw Award, the ASEE's highest recognition for contributions to technical education. The occasion marked the first time that this award was presented to two persons from the same institution, Harry Hammond having received it in 1950.

In addition to the introduction of the associate degree program, the 1953–54 academic year witnessed the creation of another curricular innovation in the form of an undergraduate course in engineering science. Dean Walker shared the view espoused by his predecessor that more Penn State engineering students should undertake graduate work. Hammond's goal of having 30 to 40 percent of the senior class go on to graduate school had never been met. Walker was especially concerned that of those students who did elect to begin graduate studies, relatively few won admission to the nation's very best engineering graduate schools, which demanded a level of theoretical expertise that most students were unable to acquire in the existing Penn State curriculums. To correct this deficiency, Walker directed that a new course of instruction—engineering science—be established to give undergraduates more extensive preparation in mathematics, physics, and other predominantly theoretical subjects. Although the inclusion of scientific subjects in engineering education became common during the 1950s, few institutions maintained separate and distinct engineering science curriculums. Penn State had one of only four such curriculums in the nation in 1959, while seven more schools offered similar programs under the title of engineering physics.[17] Engineering science at Penn State was an honors course, open only to the most capable students, and for this reason it remained a smaller program than many of its counterparts at other universities. The various departments submitted the names of their most promising freshmen, who were then personally interviewed by the dean or his surrogate before final selections were made. About fifty students per year were invited to join and were expected to maintain a minimum grade point average of 2.5 during their three years in the program. Since engineering science was not a regular academic department, it required only modest additional expenditures, and all instructors were carefully chosen from the existing departments. John W. Breneman, Professor of Mechanical Engineering, headed the curriculum until 1955, when he was succeeded by Dr. Warren E. Wilson, former president of the South Dakota School of Mines.

In conjunction with his duties as head of engineering science, Wilson was also named George Westinghouse Professor of Engineering Education, a new position funded by the Westinghouse Educational Foundation. (This appointment should not be confused with the George Westinghouse Professorship in Production Engineering, which the Foundation supported for five years in the Department of Industrial Engineering beginning in 1946.) Dean Walker realized that the quality of instruction in the College as a whole could be significantly improved. "The average engineering teacher knows little or nothing about teaching. Most engineering teachers prepare themselves for their work by getting a Master's or Doctor's degree in a professional or scientific discipline or by working in industry," he wrote in a memorandum on the subject in 1952. Yet by taking such advanced work, "which carries the teacher further from the stage of development he attained as a student, he recognizes less clearly and less easily the difficulties students have in understanding engineering principles." Walker subsequently won the interest of the Westinghouse Educational Foundation in a program to develop more effective ways of teaching engineering essentials. The

George Westinghouse Professorship in Engineering Education was the result. Professor Wilson was to study the educational methods then in use, particularly in the engineering science curriculum, and make recommendations for their improvement.

The engineering science program was an outstanding success. Virtually all the students who participated in it successfully completed the degree requirements and went on to graduate school or directly into research and development work for government or private industry. In 1974 it merged with the Department of Engineering Mechanics, which by then had developed an undergraduate curriculum of its own and was offering a course of studies in many ways paralleling that of engineering science. The engineering science curriculum of the new Department of Engineering Science and Mechanics, headed by Dr. John R. Mentzer, who had succeeded Warren Wilson as head of engineering science in 1957, remained among the most challenging in the College and continued to enjoy its status as an honors program.

The creation of the engineering science curriculum symbolized many of the changes occurring in engineering education during the post-war period. Although referring specifically only to his own field of aeronautical engineering, Professor David Peery accurately summarized these changes in 1953 in a report to Dean Walker. "Gradual modification of the subject matter in the course is required because of technical progress," observed Peery. "The long-term trend shows that the engineering profession is becoming a science rather than an art. Engineering curriculums must continually place more emphasis on fundamental scientific principles, and less emphasis on specific applications. The engineering graduate must learn the art of engineering on the job."[18] Peery had in effect described the difference between the "old" engineering education as practiced by Robert L. Sackett, John Price Jackson, and Louis Reber, and the "new" engineering education of the post-World War II era. Harry Hammond was a transitional figure in this evolution. Unlike many of his contemporaries, Hammond recognized the value of scientific studies for engineers, and under his direction, the College of Engineering slowly began to shift its emphasis to the study of theories and principles rather than practical applications. But Hammond did not go beyond taking the first few steps in this direction, his most important action being to select younger faculty who were trained in scientific research and were willing to innovate. Until sufficient numbers of these men were recruited, the undergraduate curriculums of Penn State's College of Engineering, like those of most other land-grant institutions, would continue to stress specific, job-related knowledge. In that respect they hardly differed from the curriculums of the 1920s and even earlier, prompting many of the faculty newly arrived after World War II from science-oriented institutions to consider Penn State still in the "technical institute" phase of engineering education.[19]

Eric Walker and others of the new breed of engineering educator worked hard to change the College of Engineering's academic focus. They realized that given the rapid pace of technological advance, knowledge tied to specific applications could quickly become obsolete. Consequently, devoting four years to the accumulation of such knowledge could no longer be justified, whereas a familiarity with basic scientific laws would prove valuable under any circumstance. This did not mean that the College of Engineering was about to produce masses of theoreticians. The ASEE's Grinter Report of 1955, the first detailed study of engineering education since the Hammond Reports, stated that undergraduate education must offer students the opportunity to prepare for immediate employment.[20] Yet with engineering applications growing more dependent on scientific advances (the IRL was tangible proof of this trend), science and mathematics were clearly going to comprise a larger portion of the curriculum in the years ahead. If engineering teachers understood this point, many students did not. Complaints from undergraduates that the subject matter that confronted them was too theoretical and lacking any apparent connection with practical usage became increasingly common.

Amid these major curricular modifications occurred other changes that also influenced the future of engineering at Penn State. The fine arts curriculum, which since the 1920s had been attached to the Department of Architecture, had always been something of an anomaly in the College of Engineering. No one disputed the value of a fine arts program, yet its objectives obviously were not in harmony with those of engineering education. This dissonance was aggravated in 1953 when a baccalaureate curriculum in applied arts, consisting of courses in commercial art, interior decoration, costume design, and painting and illustration, was established within the Department of Architecture. Much discussion occurred after the war regarding the future of the arts at Penn State, and in 1955 the trustees finally assented to the creation of a School of Fine and Applied Arts within the College of Liberal Arts. The Departments of Art, Music, and Theatre Arts comprised the new school, while an affiliated Department of Art Education was formed within the College of Education. Fine and applied arts instruction previously carried on under the auspices of the College of Engineering was transferred to these new divisions.

If the College thus lost one of its administrative responsibilities, it had already gained another. In September 1954, the Department of Agricultural Engineering, formerly within the province of the College of Agriculture, came under joint administration by that College and the College of Engineering as a means of securing accreditation of the agricultural engineering curriculum. The Department of Agricultural Engineering traced its birth to 1904–05, when the School of Agriculture introduced a lecture course in agricultural engineering as a requirement for all seniors. According to the Penn State catalog, topics covered in the lectures included "drainage, mechanics of machinery, building materials, the construction of substantial farm buildings, tools and implements, road-making, etc." There were no separate departments within the School until 1908, at which time the new Department of Agronomy absorbed the single agricultural engineering course and expanded it into several elective courses, such as "Farm Power and Machinery," "Farm Motors," and "Farm Sanitation." Ralph U. Blasingame, who had studied civil engineering at Alabama Polytechnic Institute and agricultural engineering at Iowa State College, became Penn State's first full-time instructor in agricultural engineering in 1913. In 1920, recognizing the importance of farm mechanization in modern agriculture, the board of trustees approved the establishment of the Department of Farm Machinery, which took over all agricultural engineering courses from the Department of Agronomy, and Professor Blasingame was named head. Like the School of Engineering's Department of Mechanics and Materials of Construction, the Department

Agricultural engineering students observing mechanical harvesting of potatoes, about 1934. (Penn State Collection)

of Farm Machinery only provided instruction for students of other departments and did not have a degree program of its own.

This situation changed in 1930, when the trustees approved the introduction of a baccalaureate curriculum in agricultural engineering. Dean Ralph L. Watts of the School of Agriculture had been campaigning for such a curriculum since the mid-1920s, arguing that most other prominent agricultural institutions were already awarding degrees in agricultural engineering. In September 1931, the Department of Agricultural Engineering (which supplanted the old Department of Farm Machinery) admitted its first class of undergraduates. Enrollment climbed steadily in the ensuing years, so that by 1954 there were 65 students and 11 faculty members. Major fields of study included land reclamation, rural electrification, farm structures, and farm power and machinery. Faculty conducted research in all these areas, with Professor John E. Nicholas's investigations in farm refrigeration perhaps being the most widely published. In 1940 the department moved into its own quarters, the Agricultural Engineering Building, located along Shortlidge Road in the northeast sector of the campus.

After Blasingame retired in 1950, another veteran agricultural engineering professor, Arthur W. Clyde, served as acting head pending the appointment of a permanent successor. Besides finding a department head, Dean Lyman E. Jackson of the School of Agriculture was faced with the necessity of having the Engineers' Council for Professional Development accredit the department's curriculum. Accreditation would permit graduates to secure professional registration as well as promote the legitimacy of agricultural engineering as a distinct field within the engineering profession. However, the ECPD refused to grant accreditation unless the department were an adjunct of the School (and later College) of Engineering. Both Harry Hammond and Eric Walker welcomed the addition of agricultural engineering, never demanding that the School of Agriculture relinquish all ties with the department. A precedent for joint administration had already been set by major midwestern universities such as Purdue, Michigan State, and Iowa State. Dean Jackson, too, saw the wisdom of joint administration, as evidenced by his selection in 1954 of Frank W. Peikert to be the next head of the Department of Agricultural Engineering. Peikert came to Penn State from the University of Maine, where he had been active in planning a similar joint administrative arrangement for agricultural engineering.

At Penn State, Peikert gave ECPD accreditation first priority. In September 1954, through a "memorandum of understanding" signed by Deans Walker and Jackson, the Department of Agricultural Engi-

Demonstration of a new mechanical egg packer and poultry house elevator, designed by agricultural engineers to save time and labor for small farmers. (Penn State Collection)

neering became a member in full standing of the College of Engineering, while retaining the same status in the College of Agriculture. Faculty and students belonged to both colleges, with Agriculture providing and administering the department's budget and Engineering granting the degrees. In 1955 ECPD representatives evaluated the department's facilities and curriculum and the following year granted a five-year accreditation. Peikert next turned his attention to expanding the graduate program and the department's physical resources. Prior to 1954, only twelve master's degrees had been awarded and no doctoral program existed. Over the next decade, 38 students earned master's degrees, and in 1968 the department inaugurated a series of courses leading to the Ph.D. That same year, after a long period of planning, a major addition to the Agricultural Engineering Building was completed. Although it did not have an associate degree program, the department did institute in 1956 a two-year course in Farm Equipment Service and Sales. Before declining enrollment forced its termination in 1976, over 300 students completed the course in preparation for employment by farm equipment dealers. In 1963, the department won approval for a separate baccalaureate major in agricultural mechanization, in which students took a variety of courses in agricultural engineering, general engineering, and business administration. Allowing students maximum flexibility in their selection of specific courses, the agricultural mechanization curriculum trained personnel for positions with equipment manufacturers and agricultural cooperatives.[21]

Perhaps the most significant administrative change in engineering occurred in May 1956, when at the personal request of Milton Eisenhower, Dr. Walker gave up his post as Dean of Engineering to assume the responsibilities of the newly created office of Vice-President for Research. Walker and Eisenhower had developed a close personal friendship, and the President, impressed by the Dean's accomplishments in advancing engineering research, hoped to take advantage of his expertise for the benefit of the entire university. Only a few weeks after Walker began his new job, however, President Eisenhower, citing "personal reasons," submitted his resignation to the board of trustees, to take effect no later than December 31. While he claimed to have been considering his decision for at least a month, Eisenhower surprised nearly everyone. In contrast to the long interregnum that followed Ralph Hetzel's death in 1947, the board of trustees lost no time in picking a successor. On October 1, 1956, Eric Walker, who carried Eisenhower's personal endorsement and who was a logical choice by any standard, became the twelfth president of The Pennsylvania State University. The new chief executive did not put aside his interest in engineering education, although he made every effort to give his successor as dean a free hand and took pains to avoid showing favoritism toward his old college. Walker continued his involvement in national engineering affairs, serving as a participant in and leader of numerous conferences and task forces. He chaired the American Society for Engineering Education committee, for example, that in 1968 issued *Final Report: Goals of Engineering Education* (the Walker Report), which followed the Grinter Report as a periodic assessment of the profession. In 1960 he was elected to a one-year term as president of the ASEE and served as president of the National Academy of Engineering (of which he was a founding member) in 1967–68. He received the ASEE's Lamme Award in 1965 and three years later was named along with Harry Hammond and twenty other distinguished engineering educators to the Society's Seventh-fifth Anniversary Hall of Fame. Nonetheless, Dr. Walker's most lasting accomplishment in the field of engineering education remained the transformation of Penn State's College of Engineering into a modern institution that won widespread recognition not only for its traditional strengths in undergraduate instruction but also for the improvements it had made in its research and graduate programs and its pioneer work in the training of engineering technicians.

6 Adapting to Changing Demands: 1956–81

The search for a new dean for the College of Engineering and Architecture commenced long before Eric Walker moved into the president's office, for Dr. Walker had tentatively accepted Milton Eisenhower's offer to become Vice-President for Research several months before he actually occupied that post. Early in 1956, at a luncheon meeting of the National Conference on the Administration of Research, Walker quite by chance was seated next to Merritt A. Williamson, then manager of the Research Division of the Burroughs Corporation, with headquarters in Paoli, Pennsylvania. The two men were already acquainted with one another. Walker informed his colleague of the impending vacancy in the deanship at Penn State and suggested that he apply. "I debated for several weeks and finally decided to respond," Dr. Williamson recalled many years afterward. "Leaving Burroughs meant a cut in salary, uprooting my wife and six children, as well as leaving behind many congenial friends and associates. However, the challenge of doing something new and different was more than I could possibly pass up."[1] Having Walker's personal recommendation, Williamson received approval from Milton Eisenhower and the trustees and in July 1956 arrived in State College to take up his new duties. (Earl Stavely had been acting dean in the interim.)

"Doing something new and different" summarized precisely the tasks awaiting the new dean. Unlike his predecessors, Merritt Williamson had virtually no experience as an educator. Just forty years old when he became dean, Dr. Williamson had devoted his entire career to the practice of engineering and administration. A native of New Hamp-

Merritt A. Williamson (Penn State Collection)

shire, he earned his Bachelor of Engineering and Master of Science degrees from Yale in 1938 and 1940 and was working on his doctorate there when commissioned in the Navy in 1944. The Navy assigned him to the California Institute of Technology, where he received a Master of Science in aeronautics and finished the research he had begun at Yale. He was awarded a Ph.D. in metallurgy from Yale in 1946. After the war, Williamson channeled his energies into the administration of research, serving successively as director of technical research for the Solar Aircraft Company in San Diego, associate director of development for the Pullman Standard Car Manufacturing Company in Hammond, Indiana, and beginning in 1952, manager of the research division at Burroughs. He also found time to obtain a Master of Business Administration degree from the University of Chicago. At Burroughs, Williamson supervised research and development in such diverse fields as physics, chemistry, metallurgy, ceramics, electronics, and materials testing. His only academic experience came in 1954, when he organized and taught (for the next two years) an evening course on the administration of research at the Moore School of the University of Pennsylvania. He soon found working in the classroom to be more rewarding and enjoyable than his regular job with Burroughs, a discovery that weighed heavily in his decision to come to Penn State.

Table 2. ENROLLMENT BY DEPARTMENT AT UNIVERSITY PARK, 1956–57

Department	Undergraduate	Graduate
Aeronautical Engineering	287	11
Agricultural Engineering	67	8
Architecture	110	0
Architectural Engineering	67	0
Civil Engineering	377	11
Electrical Engineering	941	41
Engineering Mechanics	*	27
Engineering Science	64	*
Industrial Engineering	355	3
Mechanical Engineering	591	25
Sanitary Engineering	32	3
Total	2891	129

* No degree offered in this program at this level.

Emphasis on Research

By 1956 the College of Engineering had grown to such proportions that it demanded as dean an individual who possessed superior administrative abilities, and in this respect Merritt Williamson was an appropriate choice for the post. In 1956–57, total fulltime enrollment in the College stood at 4736, more students than had attended the entire Pennsylvania State College as late as 1934. Over 3000 undergraduates and graduates were in residence at University Park, distributed among the individual curriculums as shown in Table 2. Fulltime faculty numbered 288. The College had an annual research budget of nearly $800,000 (excluding the ORL), supporting approximately fifty projects. Engineering research at Penn State had made great strides over the last decade, yet Dean Williamson saw considerable room for improvement. "There is not enough research being done by our staff," he wrote in his report to President Walker in 1957. "Those who are doing research are, in general, doing a creditable job, but there should be many more persons engaged in this activity."[2] He hoped to build on the base Walker had established earlier primarily by diversifying the nature of the research work, which in turn would attract more graduate students to the College.

The dean took three fundamental actions to stimulate research.

First, he said that the undergraduate engineering population at University Park must be held steady in order to give the faculty there more time to engage in research. Increases in enrollment must be channeled to the undergraduate centers. Next, he appointed Paul Ebaugh, formerly assistant director of the Department of Engineering Research and one of the original members of the ORL faculty, to the newly created post of Assistant Dean for Research. Ebaugh was assigned responsibilities for promoting and coordinating research throughout the College. Finally, in a related action, Dean Williamson reorganized the Department of Engineering Research into a more streamlined Engineering Experiment Department. The new department retained only the thermal laboratories and the Institutional Advisory Service under its administrative control. (Through the Advisory Service, established in 1947, the College acted as consultants to the Commonwealth's schools, hospitals, prisons, and similar installations on such matters as heating, ventilation, and insulation.) All faculty who formerly belonged to the Department of Engineering Research and who were not involved with either thermal research or the Institutional Advisory Service were transferred to the departments most closely related to their research interests. In this way, for example, the Department of Mechanical Engineering gained the services of Professors Wolfgang E. Meyer and A. W. Hussmann, two internationally respected authorities in the field of internal combustion engineering. Given Penn State's long tradition of concentrating nearly all its engineering research under one departmental authority, the Dean's plan of reorganization naturally caused some faculty members to be fearful of the changes and in some cases protest vehemently against them. Dr. Williamson persevered, nonetheless, firmly convinced that his were the best methods of encouraging more—and more varied—research, among *all* departments, not just a few.

During the next decade, the number of research projects multiplied so rapidly and reflected such diversity that it is impossible to summarize them and still do justice to the individuals and departments involved. It must be sufficient to say that the College's research budget topped the million dollar mark in 1960–61 when 72 projects were under contract. By 1965–66, research expenditures totaled $2.5 million, with 171 projects under way. Subjects investigated ran the gamut from hydraulic fluids for supersonic aircraft to mechanical egg gathering to the removal of phosphates from waste water. The federal government, through the National Science Foundation, the Atomic Energy Commission, the National Aeronautics and Space Administration, the military services, and many other agencies, regularly provided 60 to 70 percent of the funds spent on research during this period.

A major target of research dollars was the Department of Nuclear Engineering. William Breazeale had since 1954 held the title of Professor of Nuclear Engineering, and although a few graduate students had worked under his direction, no department existed in the formal sense. When in 1958 the College believed the time had finally come to organize a department, Breazeale was unable to accept the headship because physicians had advised him to seek a milder climate for reasons of health. Dean Williamson cast about for a new candidate for the post and ultimately settled on Nunzio J. Palladino, a nuclear engineer with the Westinghouse Electric Corporation. Under the guidance of Professor Palladino, who in July 1959 became the first head of the new Department of Nuclear Engineering, a formal course of studies leading to the Master of Science degree was launched. Eight students were enrolled, and in January 1961 the first degrees were awarded. The need for persons having even more specialized training soon became evident, so a doctoral program was initiated in 1963.

The resulting experimental and educational demands on the reactor facility necessitated an increase in the power level of the reactor and a corresponding enlargement of the staff. Output was boosted to 200 kilowatts in June 1960, while the General State Authority authorized funds for the construction of additional research floor space, more offices, a laboratory containing two hot cells, and other improvements. That same year the Curtiss-Wright Corporation donated to the University a 4000-kilowatt reactor at Quehanna, Pennsylvania, about fifty miles north of State College. Curtiss-Wright had endeavored to establish commercial production of isotopes at Quehanna. When the venture proved unsuccessful, the firm offered the facility to Penn State. Palladino and his colleagues at first hoped to modify the reactor for research use, but the projected expenses of the alterations combined with the remote location of the reactor forced them to abandon their plans and turn their attention to upgrading the resources already at hand. More classrooms and laboratories were added to the campus reactor facility, and the reactor itself increased in capacity to 1000 kilowatts. These improvements enabled the reactor to participate in ten times the number of research and service projects in 1965 than had been envisioned for it at the time of its construction. Much of this work was being done at the request of the electric utility industry, which had embraced nuclear power as an energy source for the generation of electricity. Utilities also required persons skilled in the design, construction, and operation of nuclear power plants. Aware of the industry's needs, the Department of Nuclear Engineering initiated a baccalaureate curriculum in 1968 and an associate degree curriculum in nuclear technology in 1970.

Nuclear reactor pool (College of Engineering)

Another important research and teaching tool that benefited both the College of Engineering and the University was PENNSTAC, Penn State's first electronic computer. ENIAC, built by the University of Pennsylvania's Moore School of Electrical Engineering in cooperation with the Army Ordnance Department and put into operation in 1946, is generally recognized as the first successful fully electronic digital computer. It and the electromechanical Mark I computer completed at Harvard in 1944 helped to encourage private firms and universities to explore the field of computer design and application. Several schools had machines in various stages of assembly in 1953 when Arthur Waynick and the staff of the Department of Electrical Engineering decided that Penn State, too, must become involved. Using $25,000 in University funds and a $17,000 grant from the National Science Foundation, Waynick's team begged and bought components from a dozen firms. The General Electric Company, having decided to discontinue most of its computer research and development, donated a complete set of plans and specifications for an electronic digital computer it had never built. After faculty and graduate students had tried and failed to build a storage drum, the International Business Machines Company made a gift of a magnetic drum it had removed from one of its older machines. In the early 1950s, few persons other than those already working with computers foresaw the wide variety of applications that these machines

PENNSTAC and control console (Penn State Collection)

were to have in the years ahead. When Waynick tried to win cooperation from the University's new College of Business in assembling and using the computer, Dean Ossian R. MacKenzie asserted, "That's a subject that won't be of any interest to business" and declined the invitation.[3] (In later years MacKenzie reversed himself and became a staunch advocate of the use of computers in business administration.) PENNSTAC, short for Penn State Automatic Computer, went into operation in July 1956. It would have cost $250,000 on the commercial market, yet it had been built at a cost to the University of less than $50,000, thanks to donations and the work of faculty and students in the Department of Electrical Engineering. PENNSTAC possessed more than monetary value to the College of Engineering, however, as its design and construction resulted in more than sixty graduate theses and the training of dozens of undergraduates for employment in the fledgling computer industry. The computer laboratory, directed by Professor Harold I. Tarpley, made its resources available to other segments of the University for research purposes. In typical applications, faculty in the College of Agriculture used PENNSTAC to assist in determining the optimum mixture for cattle feed, while chemical engineers used it to solve complex equations. Compared to computers introduced by 1960, however, PENNSTAC, with its 1500 vacuum tubes, proved expensive to maintain and slow to operate. It was eventually retired in favor of more modern facilities supervised by a new Department of Computer Science in the College of Science. The College of Engineering itself did not pursue the initiative it had gained with PENNSTAC, and by the mid-1960s its activities in design and applications lagged far behind those of many other large universities.

The Department of Electrical Engineering was also indirectly responsible for the creation of yet another department in the College of Science. In 1962 a research and graduate program in radio astronomy was begun in the Ionosphere Research Laboratory. Heading the program was Dr. John P. Hagen, former director of the National Aeronautics and Space Administration's Project Vanguard, America's first orbital space venture. The National Science Foundation provided funds for a radio astronomy laboratory, replete with a 30-foot radio telescope and other advanced equipment, which was used to correlate solar activities with changes in the earth's upper atmosphere. In 1965 the College of Science established a Department of Astronomy with Dr. Hagen as head, and work in radio astronomy was shifted from the IRL to the new department.

While perhaps lacking the glamour associated with nuclear engineering, computers, radio astronomy, and similar undertakings, another research project of the College of Engineering produced results equally as significant. Dean Williamson was even less inclined than Eric Walker to accept chronic student complaints about poor teaching as a fact of academic life. In 1957 he appointed Dr. Otis E. Lancaster to succeed Warren Wilson as the George Westinghouse Professor of Engineering Education. The George Westinghouse professorship was unique in that it was the only one in the entire country that permitted the appointee to devote all his time to the research and study of effective methods of teaching engineering. Thus far, the full potential of the position had not been realized. Dr. Lancaster, who held degrees in mathematics from the University of Missouri and Harvard University and in aeronautical engineering from the California Institute of Technology, came to Penn State from the Navy's Bureau of Aeronautics, where he had served as assistant director of the research division. He had acquired teaching experience as a member of the mathematics faculties at Harvard and the University of Maryland. Beginning in the fall of 1957, Lancaster directed annual seminars designed to improve the instructional techniques of new faculty members. A short time later he devised a questionnaire that was submitted to graduating seniors in order to elicit their opinions of the strengths and weaknesses of undergraduate instruction. In 1960 the College sponsored the first of several annual Summer Institutes on Effective Teaching for Engineering Teachers, again supervised by Dr. Lancaster and attended by engineering educators from throughout the United States and Canada. After having served for nearly a decade in the

George Westinghouse Professorship, Lancaster was named associate dean of the College of Engineering. In 1977, in recognition of his pioneering contributions to the improvement of his profession, he was elected president of the American Society for Engineering Education, thus becoming the fifth Penn State dean to have achieved that distinction. (Merritt Williamson became the fourth when he was elected ASEE president in 1969, but by that time he was no longer connected with the University.)

Money for research and curricular improvements was easier to obtain during the late 1950s and 1960s than during any previous time. A rising standard of living, the need to accommodate the educational demands of the maturing offspring of the post-war "baby boom," the increased social prestige of a college degree—these and many other factors caused the Commonwealth and the federal government to invest millions of dollars in Penn State to help it meet the challenges of the 1970s and beyond. The Soviet-American race for technological leadership that began after the USSR launched Sputnik in 1958 was especially instrumental in increasing public support for scientific and technical education. The most tangible evidence of this support could be seen in the expansion of the physical plant of the College of Engineering and Architecture. Even before Eric Walker had departed, planning was already underway for the construction of a new general engineering building, enlargement of the Main Engineering Building, and renovation of the engineering units. In January 1957, at the suggestion of Dr. Walker, the board of trustees christened the new structure (still in the design stage) Hammond Building, while the Main Engineering Building was renamed Sackett Building. Early in 1958 the General State Authority approved all construction plans and awarded contracts. The architectural firm of Howell Lewis Shay and Associates of Philadelphia designed the $5.9 million Hammond Building, which was to house the dean's offices as well as departmental offices, an expanded library, and a variety of classrooms and laboratories. A stone and metal panel structure some 600 feet long, it occupied the narrow strip of land between the engineering units and College Avenue. Initially, three buildings were to be erected there, but in an attempt to reduce expenses the three were consolidated into a single building. Much of the internal layout remained unaltered, however, so in many ways the new Hammond Building did resemble three distinct structures and presented some peculiar obstacles to students and faculty who had to negotiate their way from one end of the building to the other, particularly after business hours. Fortunately, later modifications, such as the opening of a main entrance through the new Bayard

D. Kunkle Student Activities Center at the east end of the building, eliminated most of these difficulties. To make way for the Hammond Building, the old campus power plant, for nearly thirty years the home of diesel research and the Petroleum Refining Laboratory, was demolished. Most of the diesel and other internal combustion equipment was moved to the Mechanical Engineering Laboratory (or Mechanical Engineering Building, as it is now known), but this field of research had lost much of its earlier importance. Professors Schweitzer and DeJuhasz had retired and the younger mechanical engineering faculty had developed other kinds of research interests. The Petroleum Refining Laboratory, which likewise had already experienced its years of greatest glory, was transferred to the new Chemical Engineering Building (later Fenske Laboratory), the first phase of which was completed in 1960. Construction of the north and south wings of Sackett Building was finished in 1959. Composed of the same kind of inexpensive panels used in Hammond Building, these wings bore no resemblance to what the original architects had envisioned in 1928 and did nothing to enhance the structure's artistic appeal. Renovation of the units, including the elimination of the skylighted roofs, began in 1960. In September of that year, dedication ceremonies presided over by President Walker, Dean Williamson, and Mrs. Harry Hammond marked the official opening of the Hammond Building. Four years later, yet another structure—the Electrical Engineering Building East—further increased the dimensions of the College's physical plant.

Restructuring the College

Merritt Williamson was unlike his predecessors not only in his background but in his approach to administering the College of Engineering. Whereas Sackett, Hammond, and Walker all had kept tight personal rein on the decision-making process and had acted (benignly) in semi-dictatorial fashion, Williamson was a firm advocate of the delegation of authority. At first, he had no alternative but to share some of that authority, for Governor Leader prevailed upon him in the fall of 1956 to become on a temporary basis the chairman of the politically troubled Pennsylvania Turnpike Commission. For the next eight months, the harried dean worked 80-hour weeks, attending to Commission duties three or four days a week and to affairs in the College during the remainder of the time, including weekends. Dr. Williamson's experiences in industry, together with the mounting size and complexity of the bureaucracy of the College, convinced him of the need for subordinates who could concentrate their attention on specialized areas. Thus, where Eric Walker had two assistant deans in 1956, Williamson had three assistant deans and an associate dean by 1958. Several more fundamental administrative reorganizations occurred that permanently altered the composition of the College. In 1959, a Department of General Engineering was formed, headed by Professor of Electrical Engineering Francis T. Hall, Jr. This department assumed authority over all associate degree programs, all baccalaureate instruction in engineering at the branch campuses, and instruction for the freshman year at the University Park campus. The Department of Engineering Extension was abolished, being superseded by a slimmed down Department of Continuing Education in Engineering, headed by Robert E. McCord, who had been a long-time associate of Professors Holderman and E.L. Keller. The continuing education department had charge of correspondence and short courses, informal training programs, certificate and diploma courses, and all off-campus graduate studies. About 10,000 students were enrolled at that time in the department's various programs. Dean Williamson hoped that these changes, which were made in conjunction with the formation of the University's Commonwealth Campus system, would instill within the students and faculty at the branches a greater sense of loyalty to the College of Engineering and increased participation in its activities. In a further effort to integrate continuing education with the rest of the College, the Department of Continuing Education was eliminated altogether in 1964. Professor McCord became Assistant Dean for Continuing Education and the faculty were assigned to the academic departments as continuing education specialists.

Two new organizations outside the official administrative structure of the College appeared just as engineering extension began to experience its metamorphosis. The first was the Industrial and Professional Advisory Council, the concept for which originated with Dean Williamson personally. He proposed it as an external advisory body to the College of Engineering, acting in much the same manner as the old Industrial Conferences had in the early 1920s. Each division of IPAC corresponded to one of the departments within the College. Representatives for the divisions were recommended by the departments and invited by the Dean from the ranks of professional engineers and executives of firms employing engineers. Beginning in 1959, IPAC met annually at University Park, suggesting to the Dean and the department heads ways of keeping the curriculum and research programs attuned to the needs of industry and the professions and helping to meet the needs of the College generally. Strong support from IPAC, for example, played a key role in persuading the University administration and the Commonwealth to end many years of delay and approve construction of a new addition to the Agricultural Engineering Building.

The second organization, the Penn State Engineering Society, was founded in 1958 by a group of alumni led by Michael Baker, Jr. ('36, civil engineering, and later president of the University's board of trustees). Dean Williamson had called attention to the need for such a body and applauded its formation, reasoning that it "gave the engineering alumni something to rally around besides the football team."[4] The PSES (or The Pennsylvania State University Engineering Association, as it called itself until 1966) is comprised of all College of Engineering alumni who are members of the University's Alumni Association, of which the Engineering Society is an affiliate. One of its most outstanding accomplishments has been to sponsor a fund in the College as a repository for the thousands of dollars donated annually by engineering alumni. In turn, these monies are used, with the advisory approval of the Society's board of directors, to underwrite awards for excellence in teaching, student advising, and research, as well as to support honorary engineering lectures and distinguished professorships. Part of the fund goes directly to the student body, in the form of emergency loans, scholarships, and subsidies for the *Penn State Engineer*. PSES also helps purchase books for the engineering library and disseminate information about careers in engineering to prospective students. The interest of its alumni has allowed the College of Engineering to improve the quality of its educational and social activities in ways that ordinarily would have been beyond its financial reach, even in the relatively prosperous times of the 1960s. Surely Robert Sackett and Harry Hammond would have been delighted to have had such assistance in an earlier era.

Probably the event receiving the most public attention during Merritt Williamson's deanship came in the fall of 1962. At its September meeting, the University's board of trustees acted favorably on President Walker's recommendation to detach the Department of Architecture and align it with the School of Fine and Applied Arts to create the College of Arts and Architecture, effective January 1, 1963. A new Department of Architectural Engineering, headed by Gifford H. Albright ('52, architectural engineering) was formed in what hereafter was called simply the College of Engineering. Ironically, while Dean Williamson had instigated or encouraged practically all the changes that had previously taken place within the College, he could muster little enthusiasm for this action. He feared that separating the architects and the architectural engineers would only intensify the traditional rivalry between the two groups and would reduce the effectiveness of each department. For several years the Dean had been planning the formation of a separate School of Architecture, encompassing existing curriculums in architecture, architectural engineering, and landscape architecture, and a new curriculum in industrial design. The school, which Williamson hoped would eventually be elevated to the status of a college, was to be closely associated with the fine arts program, yet would still emphasize research and instruction and otherwise prepare students for the practice of their professions.

The concept of a School of Architecture was more than just a dream. In 1958, Dean Williamson had brought Professor Albright to the University to strengthen the architectural engineering curriculum so that one day it could stand on its own as a full-fledged department. That objective was a very bold one, considering that in 1950 the American Institute of Architects had passed a resolution calling for the gradual phasing out of independent curriculums in architectural engineering. Moreover, by the time the trustees authorized the creation of the new College of Arts and Architecture, a building to house a portion of Williamson's proposed School of Architecture had already been placed on the GSA construction schedule. (The structure ultimately became part of Arts and Architecture.)

Although the administration in the end reacted unfavorably to Dean Williamson's plans, President Walker did not have to be convinced of the value of strong technical preparation for architectural students; hence the establishment of the Department of Architectural Engineering. Dean Williamson also worked to make certain that the Departments of Architecture and Architectural Engineering would work in close cooperation by having students in each curriculum take courses in the other department and appointing faculty to joint positions. The new arrangement also prevented one field from dominating the other, since each now had a separate academic home. The Department of Architectural Engineering grew rapidly from the start. By the end of the 1970s, as other schools eliminated their programs in response to the AIA resolution, Penn State had one of only eight accredited architectural engineering curriculums in the nation, enrolling more students at both the undergraduate and graduate level than any other institution. In 1964, when the Engineering Experiment Department was reorganized as the Institute for Building Research, its faculty and that of the Department of Architectural Engineering developed joint interests on many research projects. Four years later, upon the retirement of director Elmer Queer, the Institute and the Institutional Engineering Service (formerly the Institutional Advisory Service) were absorbed by the Department of Architectural Engineering, thus eliminating the last organizational vestige of the old Engineering Experiment Station.

Architectural engineering had been a component of the School and later the College of Engineering since its origin as a separate curriculum. That was not the case, however, with chemical engineering, which did not come under the administrative authority of the College of Engineering until July 1, 1963. On that date, the College of Chemistry and Physics, to which the Department of Chemical Engineering belonged, was superseded by a new College of Science. The Walker administration had earlier proposed that the College of Science be more service oriented than its predecessor—that is, it should offer many courses to students from various majors both in and out of the College. For this reason, the Department of Mathematics was detached from the College of Liberal Arts and added to Science. The Department of Chemical Engineering, on the other hand, did very specialized work and attracted few students from other departments. The faculty of the department therefore voted to join the College of Engineering.

Chemical engineering at Penn State had its foundation in the industrial chemistry curriculum, begun in 1902 in the Department of Chemistry. The course in industrial chemistry, stated the College catalog, "aims to train students primarily to be chemists . . . but with training in engineering subjects to such an extent that they will be better equipped to accept some of the promotions that come within their reach than are the graduates of the regular course in chemistry." Jesse B. Churchill was the first instructor in industrial chemistry and for several years was the School of Natural Science's only faculty member in that field. In 1924, upon formation of the School of Chemistry and Physics and in line with a national trend, the name of the curriculum was changed to chemical engineering, although it remained under the jurisdiction of the Department of Chemistry and academic requirements stayed the same. The chemical engineering laboratories and faculty offices were located in what became known in 1948 as Walker Laboratory, a red brick building razed in 1968 to make way for the present Davey Laboratory. The older building's namesake, William H. Walker, graduated from Penn State in 1890 with a degree in chemistry and two years later received a Ph.D. from Germany's Goettingen University. He then returned to his alma mater as an instructor before leaving in 1894 to join the faculty at the Massachusetts Institute of Technology. While he had no subsequent direct association with Penn State, Walker is often referred to as "the father of chemical engineering" in tribute to the pioneering work he did in that area at MIT. At Penn State, no independent Department of Chemical Engineering existed until 1948, when the board of trustees authorized its separation from the Department of Chemistry, an action that again paralleled the development of chemical engineering at many other large universities at that time. Dr. Donald S. Cryder, a native of nearby Tyrone and a 1917 graduate of the industrial chemistry curriculum, was appointed the first head of the department. He had been in charge of the chemical engineering curriculum since 1930 and had gained fame in the late 1930s by supervising the production of the first large quantity of heavy water, which was later turned over to Dr. Harold Urey of Columbia University for use in the nation's atomic research program.

Most research in chemical engineering was carried on at the Petroleum Refining Laboratory, directed by Merrell R. Fenske. Fenske, born in Indiana, did his undergraduate work at De Pauw University before earning a doctorate in chemical engineering from MIT in 1928. Dean Frank C. Whitmore of Penn State's School of Chemistry and Physics was then in the process of broadening the research base of the school and lured Fenske to the College by promising to put him in charge of investigations into the properties of crude oil. The Pennsylvania Grade Crude Oil Association, already the sponsor of several research projects in the School of Engineering, desired to have studies done in regard to the composition of Pennsylvania crude oil and the chemical and engineering aspects of its refinement. The individual members of the Association did not have the resources to undertake such studies themselves, but as a group they could marshal the funds to support the research, if Penn State would conduct it. Fenske was eager to begin the work. In 1929 he oversaw the outfitting of a petroleum laboratory in Pond Laboratory. Cutting through floors and utilizing ventilation shafts, he and his associates erected a distillation column of sufficient height to enable them to separate the components of crude oil to a greater extent than had ever been done previously. The cramped quarters of Pond Lab soon proved unsuited for further expansion, so when the old College power plant became vacant in 1931, Fenske moved his equipment there. The Petroleum Refining Laboratory, as his operation was designated, occupied both ends of the building, where the coal bunkers and turbines formerly had been. The School of Engineering's diesel laboratory was sandwiched in-between, on the site of the old boiler room. An enclosed walkway above the diesel lab connected the two wings of the refining lab. Over the next few years, Fenske and his colleagues identified and separated the individual compounds in gasoline refined from Pennsylvania crude, open-

High-vacuum fractionating column used in studies of the chemical composition of crude oil, in the Petroleum Refining Laboratory, 1933. (College of Engineering)

ing the field for similar studies in other parts of the country. The Standard Oil Company of New Jersey soon manifested an interest in sponsoring work at the laboratory, and by World War II Esso was providing at least as much financial backing as the Penn Grade Association. During the war, the Petroleum Refining Laboratory helped satisfy the armed forces' demands for oil by developing improved methods for the high-volume production of aviation gasoline and standardized automotive lubricating oils from Middle Eastern crude.

When peace returned, researchers at the lab again undertook pioneering work, this time by further separating crude oil by molecular type. But by the mid-1960s, after the laboratory had moved most of its equipment to the new Chemical Engineering Building and was officially merged into the Department of Chemical Engineering, the monetary support from the petroleum industry gradually decreased. Pennsylvania yielded diminishing amounts of crude oil, thus lessening the Penn Grade Association's need and ability to sponsor extensive research, and large oil companies had started their own research programs. The Department of Chemical Engineering then began to seek a wider variety of topics for investigation, a search that continued into the 1970s.

Whatever the merits of diversification, it was the reputation of the Petroleum Refining Laboratory that caused large numbers of graduate students to come to Penn State to study chemical engineering. Since its

Merrell Fenske (standing, with Dorothy Quiggle to his left) discussing procedures and apparatus used in the Petroleum Refining Laboratory. (College of Engineering)

founding, the Department of Chemical Engineering has regularly ranked among the nation's top ten in graduate enrollment. The numbers of undergraduates enrolled has traditionally been high, too, with the department standing seventh among 38 chemical engineering departments or curriculums in 1948, and second among 119 thirty years later. The department can also lay claim to having had the first female faculty member (as well as graduate student) of any department that currently composes the College of Engineering. Dorothy Quiggle came to Penn State from MIT with Merrell Fenske and in 1936 was awarded a Ph.D. in chemical engineering. She worked at the Petroleum Refining Laboratory throughout its existence, handling much of the administrative burden of the research. The first woman to receive a bachelor's degree from the department was Anne Needham, who graduated in 1939. Other alumni have ranged from distinguished business executives such as William L. Slater, '07 (Vice-President of the Gulf Oil Corporation) and Robert E. Kirby, '39 (Chairman of the Board of the Westinghouse Electric Corporation) to prominent engineering educators such as Max S. Peters, '42 (Dean of Engineering, University of Colorado).[5]

In addition to absorbing the Department of Chemical Engineering, the College of Engineering in the 1960s cultivated a more harmonious relationship with the College of Mineral Industries, successor to the old School of Mines. After the School of Mines had reasserted its independence from the School of Engineering in 1906, it continued to suffer from an apparent lack of interest in mineral industries education on the part of the Commonwealth, the mining industry, and even the Penn State administration. For want of money to increase the size of its instructional staff or to upgrade its inadequate physical plant and antiquated laboratory equipment, the School of Mines remained one of the smallest in the College, having an undergraduate enrollment of just 126 in 1926. At that time the School offered instruction in four baccalaureate curriculums—mining, metallurgy, geology, and ceramics—and extension work in coal mining. Although an experiment station had been authorized in 1919, it lay dormant for lack of financial sustenance, and virtually no research was being carried on. As noted earlier, Robert Sackett sought on at least one occasion to remedy these ills by having the School of Engineering take over many of the duties of the School of Mines, but he failed to win the administration's approval for such a scheme. Conditions in the School of Mineral Industries, as it was called after 1929, did not improve substantially until the 1930s and the coming of Dean Edward Steidle. During his tenure (1929–53) the Mineral Industries and Mineral Sciences Buildings were erected, the experiment station became active, and a strong graduate program was begun. In spite of the vigorous growth it experienced after World War II, the School of Mineral Industries again faced extinction in 1951, when President Milton Eisenhower's Committee on College Reorganization recommended that mineral engineering curriculums be transferred to the School of Engineering and the remaining courses be divided between the School of Liberal Arts and a new school. Dean Steidle and his faculty registered a forceful protest against this plan with President Eisenhower, who promised that the School of Mineral Industries would remain intact and soon thereafter dissolved the committee on reorganization.[6]

That engineering education at Penn State had never, with the exception of a few years around the turn of the century, come under a single administrative authority had been a continual source of ill-feeling between the College of Engineering and its counterpart in mineral industries. Attempts by the leadership in the College of Engineering to annex many of the responsibilities for mineral industries education only intensified the sparring between the two bureaucracies. In contrast to the tendencies of some of his predecessors, Merritt Williamson made a deliberate effort to cooperate with, rather than usurp the prerogatives of, the College of Mineral Industries, which by the mid-1960s contained six undergraduate engineering curriculums: ceramic technology, fuel technology, metallurgy, mineral preparation engineering, mining engineering, and petroleum and natural gas engineering. Dean Williamson went so far as to incorporate faculty representatives from the College of Mineral Industries into the College of Engineering's executive committee and in 1963 appointed as his associate dean Howard Hartman, formerly head of the Department of Mining Engineering.

As the College was moving toward a closer relationship with the College of Mineral Industries, its relationship with the Ordnance Research Laboratory was weakening. In 1963 the Walker administration removed the ORL from the jurisdiction of the College and placed it under the direction of the University's Vice-President for Research. This action was taken principally in response to the intercollegiate and interdisciplinary nature of the research in which the Laboratory participated and to encourage the growth of non-military research. Efforts were made to involve the ORL in the economic development of Pennsylvania, for example, by making it the chief technical component of the new Commonwealth Industrial Research Corporation, founded jointly in 1963 by Penn State and the Commonwealth to promote the establishment of research and development centers throughout rural Pennsylvania. Several years later the facility's name was changed to the Applied Research Laboratory, again reflecting the University's desire to broaden the research base of the facility beyond purely defense-oriented work. Dean Williamson was reluctant to see an end to the College of Engineering's hegemony over the Laboratory but, as in the case of the formation of the College of Arts and Architecture, he had little choice but to comply with the directives of his superiors. To ensure continued cooperation with the Laboratory, however, he invited its director to retain his membership on the College of Engineering's executive committee. Many faculty also continued to hold joint appointments, and the physical resources of the ARL continued to be made available to the College's faculty and students.

Far more troubling to Williamson personally was the deadening bureaucratic routine that had crept into much of his work after nearly a decade in the dean's office. He missed the adventure of doing new things, of exploring new methods. Therefore, when in 1966 Vanderbilt University invited him to become its Orrin Henry Ingram Distinguished Professor of Engineering Management, he accepted. Ten years of being dean had instilled in Dr. Williamson a definite preference for the academic side of engineering, and this new position would not only allow him to shape a new academic program, but to teach, something he had longed to do since he had organized his evening class at the University of Pennsylvania. He submitted his resignation to President Walker in May 1966 and in August left State College to assume his new responsibilities.

The Palladino Years

Ever since Robert L. Sackett's appointment in 1915, the selection of the new dean had been made (at least informally) prior to the departure of the old. But in 1966, for the first time in half a century, a question existed as to who would be the next dean of engineering. Howard Hartman was named acting dean until a permanent replacement for Merritt Williamson could be found. The choice eventually settled on N.J. Palladino, head of the Department of Nuclear Engineering, even though he had not been among the candidates who actively sought the post. On October 1, 1966, he became the College of Engineering's seventh dean.

When Dean Williamson selected Palladino to head the Department of Nuclear Engineering in 1959, he had hailed the new appointee's "excellent background and enviable reputation in the design and construction of nuclear power plants."[7] Yet before coming to Penn State, "Joe" Palladino had never really thought of himself as a nuclear engineer. A native of Allentown, Pennsylvania, he had originally intended to fulfill a boyhood dream by going into railroad work and earned a bachelor's degree in mechanical engineering from Lehigh University with that career in mind. The lingering effects of the Depression necessitated a change in plans, however. In 1939, after completing graduate work in mechanical engineering at Lehigh, Palladino accepted a position with the Westinghouse Electric Corporation at its South Philadelphia works and was eventually assigned to the steam turbine design section. After serving in the Army during the war, he was sent by Westinghouse to Oak Ridge, Tennessee, where the federal government contemplated building a nuclear reactor and where the young mechanical engineer's expertise in the field of heat transfer would be of great value. The Oak Ridge project never materialized, but Palladino, still on loan from Westinghouse, was transferred to Argonne National Laboratory, where his knowledge of heat transfer was again put to use, this time in applications studies for nuclear-powered submarines. At both Oak Ridge and Argonne, he studied and taught courses in reactor physics. Consequently, when Westinghouse in 1949 decided to supply components for nuclear submarines, it recalled Palladino and assigned him to the company's Bettis Laboratory

Nunzio J. Palladino (College of Engineering)

in Pittsburgh. There he helped design the reactor for the *Nautilus,* the Navy's first nuclear submarine. Later he participated in designing the Shippingport (Pennsylvania) reactor, America's first commercial venture in the generation of electricity through atomic energy.

Palladino had never given much consideration to an academic career in engineering, despite the occasional teaching he was doing for Westinghouse. ("Imagine me a professor!" was his initial reaction to a query from Dean Williamson about the headship of the Department of Nuclear Engineering.)[8] Still, he enjoyed interacting with students and the challenges of the classroom, and, since Westinghouse was not giving him as much opportunity as he preferred to work in the commercial reactor field, he accepted the offer to come to Penn State. As head of nuclear engineering for seven years, he brought his teaching experience and his organizational abilities to bear on what then was a pioneer program. When he left the department in 1966, its graduate curriculum and training courses for nuclear power plant operators were among the most respected in the nation.

Dean Palladino faced some even more formidable problems in his new position. A four-year slump in undergraduate admissions sent the size of the student body tumbling by over 600 between 1969–70 and 1972–73. Engineering schools across the land experienced this same downswing, which stemmed in part from the reduction of activity in the aerospace industry. Since that industry employed not only aerospace engineers but large numbers of personnel in the electrical, mechanical, and other engineering fields also, layoffs by the federal government and its contractors had widespread repercussions. Exacerbating the problem were the predictions of many so-called experts, who, panicking at the decline in aerospace and related areas, foresaw a less than prosperous future for the engineering profession in general. This pessimism surfaced despite the findings of the American Society for Engineering Education's Committee on the Goals of Engineering Education, chaired by Eric Walker. After a careful assessment of current and future national needs, this committee in 1968 projected that the demand for engineers would rise significantly during the next decade. The committee urged engineering educators and the engineering profession "to make every effort to attract an increasing number of students to engineering at both the undergraduate and graduate level" and called for "a continuation and increased development of technician programs."[9] But attracting more students to engineering was difficult for Penn State and other institutions in the early 1970s, as many young people, seemingly alienated by technology and materialism, turned to careers that they believed to be more humanistic and idealistic. Many of those who were technically inclined chose science rather than engineering, reflecting the fact that since World War II, scientists had been winning an increasing amount of the public esteem that in earlier generations was reserved for engineers. The intensifying American military involvement in Southeast Asia and the campus unrest that appeared at home spurred on this trend. The University escaped most of the disruptions that plagued other large campuses, although the Applied Research Laboratory did become a favorite target of disgruntled students and other activists. The College of Engineering, too, came under fire briefly from anti-war protestors, mainly because of the large amounts of defense-related research allegedly being conducted there and the College's link, however tenuous, with military contractors. Demonstrations reached their zenith early in 1970. The situation seemed so threatening at one point that, for several days, the College's faculty and staff did nightly patrol duty in most of the engineering buildings in order to discourage property damage.

As the Vietnam War wound down, the decline in baccalaureate enrollment reversed itself, thanks to an increased demand for engineers by industry and government and uncertain employment opportunities in many non-technical fields. This demand remained strong through the late 1970s, and by 1980–81, with 7056 undergraduates, Penn State's College of Engineering had one of the largest engineering enrollments in the nation. Even before the cessation of the Vietnam War, the College took the initiative to interest more members of racial minorities and more women in careers in engineering. For various reasons, the number of blacks entering the engineering profession traditionally had been very small. Those few blacks who did choose to become engineers usually attended colleges and universities in urban areas, close to large concentrations of the black population. Hoping to attract more blacks to engineering at Penn State and to the profession in general, a recruiting campaign headed by Assistant Dean Ernest R. Weidhaas was launched by the College in 1968. It featured advertisements in periodicals and spot announcements on radio and television and predated similar University-wide recruitment efforts by several years. The campaign did not lead to a significant change in the proportion of black engineering students at Penn State, however, although it may have contributed to a gradual increase in the overall number of blacks who are becoming engineers. A parallel attempt to bring more females into the ranks of the University's undergraduate engineering population was much more successful, possibly because women did not regard the rural setting of the institution as a serious liability. Women could expect to encounter some prejudice, of course, upon entering a field that throughout its history had been comprised almost exclusively of men. When in 1955 Eric Walker remarked that "the most evident ambition of many women is to get married and raise a family, and certainly not to pursue a career in engineering,"[10] he echoed a sentiment shared by the vast majority of his colleagues nationwide. And in truth, in 1955 very few women seriously entertained thoughts of becoming engineers. But by the late 1960s, reflecting trends in society at large, if not among fulltime practicing engineers, attitudes were beginning to change toward women in the profession. In 1967–68, of the 3714 undergraduates enrolled in the College of Engineering, just 26, or less than 1 percent, were female. Given an initial stimulus by the College's recruitment program, these numbers rose steadily in ensuing years. A notable milestone was reached in 1970–71 with the organization of the second chapter of the national engineering sorority Beta Rho Delta. (The first chapter was founded in 1967 at Arizona State University.) A branch of the Society of Women Engineers later superseded Beta Rho Delta at Penn State. By 1980–81, women comprised 14 percent of the undergraduate student body in the College of Engineering. Their distribution by curriculum and sex is shown in Table 3.

While the undergraduate population expanded, the numbers of students seeking advanced degrees in engineering fell as the 1970s wore on, both at Penn State and across the country. The ASEE's Goals Committee had recommended that more students be encouraged to undertake graduate work and in fact had taken a stronger

Table 3. UNDERGRADUATE ENROLLMENT, COLLEGE OF ENGINEERING, 1980–81

Curriculum	Undergraduate (Male)	Undergraduate (Female)	Total UG	% Female
Aerospace Engineering	183	12	195	6
Agricultural Engineering	77	7	84	8
Architectural Engineering	245	65	310	21
Chemical Engineering	379	93	472	20
Civil Engineering	351	53	404	13
Electrical Engineering	685	52	737	7
Engineering Science	93	14	107	13
Environmental Engineering	69	17	86	20
General Engineering*	3105	538	3643	15
Industrial Engineering	217	74	291	25
Mechanical Engineering	581	38	619	6
Nuclear Engineering	93	15	108	14
Total	6078	978	7056	14

*Undeclared major, mainly freshmen and sophomores.

stand in favor of graduate education than any of its predecessor committees. The question of whether the undergraduate curriculum alone should constitute an engineer's specialized professional training (as it did for architects and pharmacists, for example) or whether it should serve as a period of general education had for many years posed a dilemma for engineering educators. No ASEE panel had taken a definitive position on the issue until the mid–1960s, when the Walker report declared that one year of graduate work should be added to the baccalaureate study that had heretofore comprised the basic education that most engineers needed for entry into the profession. The recommendation evoked widespread controversy and in the years that followed remained largely unimplemented, even at the school where Eric Walker's influence had been so pervasive.

President Walker retired in 1970 after presiding over one of the greatest periods of expansion in the University's history—a time when the institution's physical plant, student enrollment, and curricular offerings grew at an unprecedented rate. He was succeeded by John W. Oswald, formerly president of the University of Kentucky and executive vice-president of the University of California system.

With few exceptions, graduate work had never been a strong point of Penn State's College of Engineering, which throughout the 1960s and 1970s remained far down the list of large graduate institutions, even though strenuous efforts were made to enlarge the size of the graduate student body without diluting the quality of scholarship. The demand for engineers at entry level positions grew so rapidly in the years after the Vietnam War that fewer students saw much economic reward in pursuing advanced studies immediately following the completion of their undergraduate education. The number of graduate students in engineering schools across the nation steadily diminished through the late 1970s, a phenomenon that posed serious problems for the United States in the 1980s and beyond, particularly since an increasing proportion of foreign students made up the shrinking graduate population. Yet as the College's graduate enrollment fell, finally stabilizing at about 430 as the decade ended, its research expenditures climbed, surpassing $5.6 million by 1979–80. Special attention was given to obtaining additional financial sponsorship of research from non-governmental sources.

The College also recognized that it must continually change and upgrade its curriculums in order to meet the changing requirements of

An engineering graphics class of the 1980s. (College of Engineering)

the profession and society. Thus a broad course of studies in environmental engineering replaced the sanitary engineering curriculum in 1970–71. The Department of Industrial Engineering, following the pattern set at many other institutions, expanded beyond its previous emphasis on production engineering to encompass such topics as process design, management science, and systems analysis. In 1973 it changed its name to the Department of Industrial and Management Systems Engineering to better reflect the diversity of its offerings. Entirely new graduate curriculums in bioengineering and acoustics were introduced as interdisciplinary programs. An engineering technology curriculum leading to the Bachelor of Technology degree was begun at the Capitol Campus near Harrisburg in 1967 and by 1979–80 offered instruction in the fields of building construction, electrical design, mechanical design, transportation, and water resources. Administered by the Capitol Campus, the program grew from an initiative within the College of Engineering and was created to meet the need for a kind of hybrid professional whose responsibilities included both those of the engineer and the engineering technician. The engineering technology program at the Capitol Campus reflected a national trend. Where only one institution had an accredited baccalaureate curriculum in engineering technology in 1967, in 1973—when Penn State's program gained accreditation—twenty-eight other institutions were offering accredited instruction leading to a bachelor's degree in engineering technology.[11]

The four-year engineering technology program did not undercut the courses of study at the associate degree level. Demand for engineering technicians remained strong, and the two-year curriculums that were introduced in the 1970s—biomedical equipment technology, railway engineering technology, solar heating technology, and telecommunications technology, to name a few—were not available in the Capitol Campus baccalaureate program. Nor did state support for community colleges, belatedly offered by the legislature in the mid-1960s, draw significant numbers of students away from Penn State's engineering technology studies. While the Commonwealth of Pennsylvania as a whole was not a national leader, Engineers Joint Council surveys reveal that Penn State led all other institutions throughout the decade of the 1970s in the number of Associate in Engineering degrees conferred annually; and by 1979, Capitol Campus was awarding the second highest number of baccalaureate engineering technology degrees in the nation. The College of Engineering's continuing education program, on the other hand, did suffer from competitive courses offered by community colleges as well as from continued emphasis on the college degree as the key to professional success. Total enrollment dropped from 27,270 in 1968–69 to 16,659 eleven years later.

On the whole, the decade of the 1970s was a time of renewed interest in engineering. However, the record numbers of undergraduates admitted sorely taxed the College's resources at a time when the University was forced to make severe budgetary retrenchments. To accommodate the growing student load, the College of Engineering had to resort—albeit with considerable reluctance—to hiring temporary instructors for introductory courses, increasing class size, and reducing faculty assignments to non-sponsored research. The College also faced the necessity of having to replace a large proportion of its laboratory gear and other hardware, much of which had been acquired around 1960, when Hammond Building was constructed. Some dated from World War II and even earlier. Inflation and the University's need to institute major economies prevented sweeping modernization. "We have about $10 million worth of equipment," Dean Palladino remarked in 1979. "Much of this equipment has a good, useful life of about five years. We should have been putting in at least a million dollars per year in the last ten years, and we haven't been able to put in 10 to 20 percent of that."[12]

Desiring to return to teaching and other non-administrative duties for the year remaining before his retirement, Palladino late in 1980 submitted his resignation as dean, effective July 1, 1981. He expected to assume the title of University Professor, a distinction awarded only to those outstanding Penn State faculty who represent the University rather than a particular department or college. This new position would allow him to lecture, write, and teach on topics of interest to him and the Penn State community. However, in the spring of 1981, President Reagan nominated him to head the Nuclear Regulatory Commission. The Senate quickly confirmed the nomination, and Palladino assumed his new duties as soon as his resignation from the deanship took effect.

Palladino's appointment to the NRC (successor to the Atomic Energy Commission) testified to the close ties he had maintained with practical matters in nuclear engineering throughout his tenure as dean. He served as president of the American Nuclear Society in 1970–71, for example, and as a member of several state and federal technical advisory bodies, including the AEC's Committee on Reactor Safeguards. As important as any factor in bringing Palladino to the attention of the White House was his work in connection with the accident at the Three Mile Island nuclear power plant near Harrisburg in March 1979. In June of that year, Pennsylvania's Governor Dick Thornburgh chose

Wilbur L. Meier (Purdue University Photo)

Palladino as one of fourteen members of a commission to investigate the accident; and in 1980, the NRC itself named him to a nine-member task force evaluating clean-up operations at the power plant.

Succeeding Palladino as dean of engineering was Dr. Wilbur L. Meier, since 1974 head of the School of Industrial Engineering at Purdue University. A 1962 graduate of the University of Texas and a former professor of industrial engineering at Texas A&M and Iowa State universities, Meier had achieved much success in elevating the stature of Purdue's industrial engineering program. The biggest challenge of his career lay ahead, however, as he assumed a position of leadership in carrying forward engineering education at The Pennsylvania State University into its second century.

Retrospect

The financial hardships experienced by the College of Engineering in the 1970s and early 1980s were in many ways similar to those under which students and faculty had labored during the time of Louis Reber and John Price Jackson. But if financial constraint has characterized the history of Penn State's College of Engineering, so, too, have high standards of instruction, research, and service. It is unfortunate, therefore, that from its founding until World War II, The Pennsylvania State University has frequently been portrayed as mainly an agricultural school. This image is derived in part from the appellation—The Farmers' High School—the institution first gave itself and from the rural setting of the University: how could any institution of higher learning set in such splendid geographic isolation be anything but a farm school, particularly in the years before high-speed transportation and communication? Reinforcing the agricultural complexion of the University is a popular misunderstanding of the aims of the Morrill Land Grant Act of 1862. This legislation is frequently remembered, possibly because of the very words "land grant," as fostering work exclusively in agricultural education and research, when in reality it directed land-grant schools to, among other things, "teach such branches of learning as are related to agriculture and the mechanic arts." Yet the image of the agricultural college has prevailed not only in the eyes of persons having a casual acquaintance with the institution, but among a substantial portion of alumni and faculty as well. Even professional historians have adopted this view of the University. Those Pennsylvania history textbooks that give more than passing reference to Penn State, for example, invariably place its development in the nineteenth and early twentieth centuries almost entirely within the context of the rise of scientific agriculture.[13] In spite of its widespread acceptance, however, this characterization of Penn State as little more than a glorified "cow college" for nearly the first hundred years of its existence is a false one and obscures a significant chapter in the history of higher education in the Commonwealth and in the history of technological education in general. From about 1890 until 1910, the engineering curriculums enrolled at least one-half of Penn State's undergraduate student body and over one-fourth until the onset of the Great Depression. That public recognition of and esteem for The Pennsylvania State University began to grow at the same time the institution introduced its first baccalaureate curriculums in engineering is not coincidental. Engineering provided Penn State with a utilitarian purpose that citizens of America's most industrialized state could appreciate, particularly since—true to the spirit of land-grant education—degrees in engineering could be earned by persons of modest financial means. However, the service that the College of Engineering performed for Pennsylvania and the nation by graduating thousands of individuals conversant with a wide variety of technical subjects did not immediately result in more generous fiscal support from the Commonwealth. Indeed, allocations from Harrisburg simply to maintain instruction in engineering, let alone improve or expand it, were wretchedly meager until the late 1920s. When the legislature and governor at last realized that support of engineering education at Penn State was in effect an

Physical plant of the College of Engineering in 1981. (College of Engineering)

investment in the future of Pennsylvania, the Great Depression intervened and assistance remained small. After World War II, federal funds for education and research allowed the College of Engineering to enter an era of relative prosperity, as it upgraded its physical resources, restructured many of its curriculums, and otherwise prepared to meet the growing demand for technological training.

This is not to say that in its earlier years the College of Engineering was unreceptive to new approaches. The historical record suggests otherwise. Penn State was the first institution in the United States to offer degree curriculums in industrial engineering (and milling engineering, a curriculum that has long been extinct), and one of the first two or three institutions to begin baccalaureate studies in sanitary engineering and architectural engineering. It was the first in Pennsylvania to have accredited curriculums in aeronautical engineering, agricultural engineering, and nuclear engineering. In the form of technical institutes and industrial training programs, the College of Engineering developed a broad course of studies in extension or continuing education long before most other baccalaureate degree-granting schools had even undertaken their initial ventures in that area. The technical institutes prepared the way for associate degree studies in engineering technology, a field in which Penn State was again a leader among baccalaureate institutions. The institutes also laid the foundation for the establishment of the University's Commonwealth Campus system by demonstrating the feasibility of maintaining branch campuses that brought higher education to Pennsylvania residents who had no other opportunity to pursue such studies.

The College of Engineering's strength historically lay with instruction rather than research. Although the Engineering Experiment Station was one of America's first academic facilities devoted to engineering research, and its work in diesel engineering and heat transfer in building materials won international recognition, concentration of most research activity in the station undermined the attempts of various departments to launch significant research programs of their own. The geographic isolation of State College also inhibited the development of research in an era before the federal government became a major financial contributor to such activity. Absence of close physical contact with industry for many years handicapped the College's efforts to attract private financial sponsors as well as first-rate research engineers and graduate students. After World War II, as more money to support research became available and the College's leadership grew more sensitive to the need for scientific investigation, the situation improved. The nuclear reactor, the Ionosphere Research Laboratory, and the Ordnance Research Laboratory are representative of the diversity that characterized the scope of research in the College since the war.

In sum, after a slow and hesitant beginning, engineering education emerged as a major function of The Pennsylvania State University and enabled the institution to take a paramount position among land-grant schools in preparing the vast numbers of persons needed to solve the nation's increasingly complex technical problems. Pennsylvania especially required a large reserve of technically trained personnel, and for nearly a century Penn State has been the Commonwealth's largest single academic source of professional engineers. Since 1884, the College of Engineering has awarded over 11,000 associate degrees, 23,000 baccalaureate degrees, and 2500 graduate degrees, in addition to having trained countless thousands of men and women through continuing education programs. In instruction, research, and service, the College has more than fulfilled the hopes and expectations of Evan Pugh, Justin Morrill, and other pioneer leaders in the land-grant movement.

Notes

In order to keep the number of notes to a useful minimum, specific citations have been given mainly for direct quotations (where the source is not identified in the text) or in cases where a substantial amount of information was drawn from a source other than the standard sources used in the preparation of this book, namely, official reports of faculty and administrators, interviews with faculty and alumni, the *Penn State Engineer* and other student publications, various alumni periodicals, the two full-length histories of the University, and biographical and miscellaneous information on file in the Penn State Room of the University's Pattee Library. *Engineering Education* and other publications of the American Society for Engineering Education constituted the chief source of information regarding national developments in engineering education.

Chapter 1

1. *The Agricultural College of Pennsylvania, Embracing a Succinct History of Agricultural Education in Europe and America, together with the Circumstances of the Origin, Rise, and Progress of the Agricultural College of Pennsylvania, etc.* (Philadelphia, 1862), p. 45.

2. George S. Emmerson, *Engineering Education: A Social History* (New York: Crane, Russak, 1973), pp. 140–53; Daniel C. Calhoun, *The American Civil Engineer: Origins and Conflict* (Cambridge: Massachusetts Institute of Technology Press, 1960), pp. 43–54, 182–99; Lawrence P. Grayson, "A Brief History of Engineering Education in the United States," *Engineering Education* 68 (December 1977): 248–50.

3. James G. McGivern, *First Hundred Years of Engineering Education in the United States, 1807–1907* (Spokane: Gonzaga University Press, 1960), p. 87; Saul Sack, *History of Higher Education in Pennsylvania*, 2 vols. (Harrisburg: Pennsylvania Historical and Museum Commission, 1963), Vol. 2, pp. 478–82.

4. Quoted in Erwin W. Runkle, "The Pennsylvania State College, 1853–1932: Interpretation and Record" (unpublished manuscript, State College, Pa., 1932), p. 152.

214   Notes

5. "Report of the Professor of Mathematics and Astronomy," in *Annual Report of The Pennsylvania State College for 1880* (Harrisburg, 1881), p. 38.

6. Sack, *Higher Education*, Vol. 2, pp. 488–90; Grayson, "Engineering Education," p. 251.

7. Quoted in Runkle, "Interpretation and Record," p. 210.

8. Osmond to E.W. Runkle, 15 October 1913, quoted in Ibid., p. 215.

9. *Report of the Committee of the General Assembly, Appointed at the Request of the Board of Trustees, to Investigate the Affairs of The Pennsylvania State College* (Harrisburg, 1883), pp. 11–13.

10. Ibid., p. 105.

11. Louis E. Reber, "Recollections of The Pennsylvania State College, 1876–1907," in Reber Author and Biographical Vertical File (ABVF), Penn State Room, Pattee Library of The Pennsylvania State University.

12. Frederick E. Terman, "History of Electrical Engineering Education in the U.S.," *Proceedings of the Institute of Electrical and Electronics Engineers* 64 (December 1976): 1399–1401.

13. "President's Report," in *Annual Report of The Pennsylvania State College for 1889* (Harrisburg, 1890), p. 11.

14. All quotations from "Addresses Delivered at the Formal Opening of the Engineering Building of The Pennsylvania State College," in *Annual Report of The Pennsylvania State College for 1893* (Harrisburg, 1894), pp. 112–23.

15. Reber, "Recollections."

16. Ibid.

17. H.L. Plants and C.A. Arents, "History of Engineering Education in the Land-Grant Movement," in *Proceedings of the Association of Land-Grant Colleges and State Universities* (Washington, 1962), pp. 88–91; Raymond Merritt, *Engineering in American Society, 1850–75* (Lexington: University Press of Kentucky, 1969), pp. 27–62; Monte Calvert, *The Mechanical Engineer in America, 1830–1910* (Baltimore: Johns Hopkins Press, 1967), pp. 43–85.

18. "Report of the Department of Civil Engineering," in *Annual Report of The Pennsylvania State College for 1892* (Harrisburg, 1893), p. 24.

Chapter 2

1. "Report of the School of Engineering," in *Annual Report of The Pennsylvania State College for 1900–01* (Harrisburg, 1901), p. 34.

2. Ibid.

3. Sack, *Higher Education*, Vol. 2, p. 471.

4. "Report of the Department of Electrical Engineering," in *Annual Report of The Pennsylvania State College for 1897* (Harrisburg, 1898), p. 44.

5. "Report of the School of Engineering," in *Annual Report of The Pennsylvania State College for 1899–1900* (Harrisburg, 1900), p. 68.

6. Quoted in Reber, "Recollections."

7. "Report of the School of Engineering," in *Annual Report of The Pennsylvania State College for 1906–1907* (Harrisburg, 1907), p. 36.

8. "Report of the Department of Mechanical Engineering," in *Annual Report of The Pennsylvania State College for 1901–02* (Harrisburg, 1902), p. 71.

9. Sack, *Higher Education*, Vol. 2, pp. 463–76; Wayland F. Dunaway, *History of The Pennsylvania State College* (Lancaster, Pa., 1946), pp. 81, 174.

10. Quoted in "Report of the School of Engineering," in *Annual Report of The Pennsylvania State College for 1901–02* (Harrisburg, 1902), p. 50.

11. Reber, "Recollections."

12. Ibid.

13. *The Engineer,* May 1908, p. 15.

14. "Report of the School of Engineering," in *Annual Report of The Pennsylvania State College for 1907–08* (Harrisburg, 1908), p. 76.

15. Charles W. Lytle, "Collegiate Courses for Management: A Comparative Study of the Business and Engineering Colleges," *Journal of Engineering Education* 32 (June 1932): 806–07.

16. "President's Report," in *Annual Report of The Pennsylvania State College for 1911–12* (Harrisburg, 1912), p. 15.

17. *Penn State Alumni Quarterly,* October 1911, p. 15; Ridge Riley, *Road to Number One* (New York: Doubleday, 1977), pp. 110–11.

18. "Autobiography," Dedrick ABVF, Penn State Room, Pennsylvania State University.

19. Plants and Arents, "Land-Grant," pp. 92–93; Grayson, "Engineering Education," p. 256; A. A. Potter, "Engineering Experiment Stations," *Bulletin of the Society for the Promotion of Engineering Education* 6 (April 1916): 615–19.

20. "Report of the Department of Electrical Engineering," in *Annual Report of The Pennsylvania State College for 1902–03* (Harrisburg, 1903), p. 53.

21. "Report of the School of Engineering," in *Annual Report of The Pennsylvania State College for 1909–10* (Harrisburg, 1910), p. 30.

22. N.C. Miller, "History of Engineering Extension at The Pennsylvania State College," *Pennsylvania State College Bulletin,* Vol. 16, No. 1; Plants and Arents, "Land-Grant," p. 93.

23. *Annual Report of The Pennsylvania State College for 1911–12* (Harrisburg, 1912), p. 15.

24. Charles R. Mann, "Report of the Joint Committee on Engineering Education," *Bulletin of the Society for the Promotion of Engineering Education* 9 (September 1918): 16–32.

25. Harding ABVF, Penn State Room, Pennsylvania State University.

26. Paul D. Simpson, "The Bellefonte Central Railroad," *Railroad and Locomotive Historical Society Bulletin* 117 (October 1967): 7–17.

Chapter 3

1. Interview with Kenneth L. Holderman, State College, Pa., 13 August 1979.

2. "Annual Report of the School of Engineering for 1915–16." Typewritten copies of the annual reports of the School (and later College) of Engineering from 1914–15 to date are filed in the office of the Dean.

3. Quoted in *Penn State Alumni Quarterly,* January 1917, p. 65.

4. *Penn State in the World War* (State College: Alumni Association of The Pennsylvania State College, 1921), pp. 16–17.

5. "Annual Report of the School of Engineering for 1918–19."

6. *Pennsylvania State College: Its Needs and Services* (Harrisburg: Pennsylvania State Chamber of Commerce, 1921), pp. 7–10, 41–42.

7. *Penn State Alumni News,* June–July 1921, p. 9.

8. Ibid., January 1922, pp. 71–73.

9. The full report of the Conference is included in "Annual Report of the School of Engineering for 1921–22."

10. "Annual Report of the President for 1922–23," *The Pennsylvania State College Bulletin,* December 1923, pp. 78–79.

11. *Penn State Engineer,* October 1924, p. 6.
12. Louis J. Venuto, "Dean Edward Steidle's Contributions to the Growth of the College of Mineral Industries at The Pennsylvania State University: A Case Study" (Ph.D. dissertation, The Pennsylvania State University, 1965), pp. 85–86.
13. *Penn State Engineer,* May 1924, p. 17.
14. "Annual Report of the School of Engineering for 1923–24."
15. Ibid.
16. Interview with Paul Schweitzer, State College, Pa., 21 August 1979.
17. Miller to Thomas, 23 November 1921, appended to "Annual Report of the School of Engineering for 1921–22."
18. "Annual Report of the School of Engineering for 1923–24."
19. "Annual Report of the School of Engineering for 1924–25."
20. "Annual Report of the President for 1926–27," *The Pennsylvania State College Bulletin,* December 1927, p. 3.

Chapter 4

1. Memorandum to the President for 1935–36, filed with "Annual Report of the School of Engineering for 1935–36."
2. Interview with Clifford B. Holt, Jr., State College, Pa., 24 September 1979.
3. Thorndike Saville, "Achievements in Engineering Education," *Journal of Engineering Education* 43 (December 1952): 222–35; C. E. Davies, "The Builders of the Engineers' Council for Professional Development," in *Twenty-fifth Annual Report of the ECPD* (New York: ECPD, 1957), pp. 25–28.
4. "Annual Report of the School of Engineering for 1936–37."
5. *Penn State Engineer,* October 1937, p. 7.
6. "Annual Report of the School of Engineering for 1937–38."
7. *Report of the Investigation of Engineering Education, 1923–29,* 2 vols. (Pittsburgh: Society for the Promotion of Engineering Education, 1934).
8. *Penn State Engineer,* February 1939, p. 16.
9. Harry P. Hammond, "Promotion of Engineering Education in the Past Forty Years," *Journal of Engineering Education* 24 (September 1933): 59.
10. "Report of the Committee on Aims and Scope of Engineering Curricula," *Journal of Engineering Education* 30 (March 1940): 555–66.
11. "Semi-Annual Report of the School of Engineering," 31 December 1938.
12. Saville, "Achievements," pp. 223–25.
13. Quoted in *Penn State Alumni News,* February 1937, p. 11.
14. "Annual Report of the School of Engineering for 1940–41."
15. A.A. Potter, "Engineering Education—The Present," *Journal of Engineering Education* 34 (September 1943): 34–40; *Penn State Engineer,* January 1942, p. 11.
16. Letter to the author from Lidia Manson, 30 August 1979.
17. "Annual Report of the School of Engineering for the Biennium, 1941–43."

Chapter 5

1. "Report of Committee on Engineering Education after the War," *Journal of Engineering Education* 34 (May 1944): 589–614.
2. Plants and Arents, "Land-Grant," pp. 95–96; Grayson, "Engineering Education," pp. 259–60.
3. Interview with Eric A. Walker, State College, Pa., 2 August 1979.

4. *Report of the President to the Trustees of the College and the People of Pennsylvania, July 1, 1945–June 30, 1947* (State College, 1947), pp. 14–15.
5. "Annual Report of the School of Engineering for 1944–45."
6. Quoted in *Penn State Engineer,* November 1949, p. 7. See also *Garfield Thomas Water Tunnel, 1949–74* (University Park: Applied Research Laboratory of The Pennsylvania State University, 1974).
7. "Annual Report of the School of Engineering for 1948–49."
8. *Penn State Engineer,* December 1953, p. 17.
9. Interview with Kenneth L. Holderman, State College, Pa., 13 August 1979.
10. "Annual Report of the Department of Mechanical Engineering for 1953–54."
11. Interview with Eric A. Walker, State College, Pa., 2 August 1979.
12. Ibid.
13. Ibid.; *Fifteenth Semiannual Report of the Atomic Energy Commission* (Washington: Government Printing Office, January 1954), p. 35; *Eighteenth Semiannual Report of the Atomic Energy Commission* (Washington: Government Printing Office, July 1955), pp. 102–3; Forrest J. Remick, "History of Nuclear Engineering at Penn State," 1967 (typewritten).
14. *Penn State Engineer,* January 1953, p. 24.
15. K.L. Holderman, J.W. Breneman, and E.A. Walker, "The Echelons of Engineering Education," *Journal of Engineering Education* 44 (December 1953): 234–42; *Twenty-first Annual Report of the Engineers' Council for Professional Development* (New York: ECPD, 1953), pp. 48–51; G. Ross Henninger, *The Technical Institute in America* (New York: McGraw-Hill, 1959), pp. 4–11.
16. Interview with Eric A. Walker, State College, Pa., 2 August 1979.
17. *Twenty-seventh Annual Report of the Engineers' Council for Professional Development* (New York: ECPD, 1959), pp. 38–39.
18. "Annual Report of the Department of Aeronautical Engineering for 1952–53."
19. Interview with Arthur H. Waynick, State College, Pa., 8 May 1980.
20. "Report of the Committee on Evaluation of Engineering Education," *Journal of Engineering Education* 46 (September 1955): 26–60.
21. Frank W. Peikert, *History of Agricultural Engineering at Penn State, 1892–1976* (University Park: College of Agriculture of The Pennsylvania State University, 1976), pp. 7–45.

Chapter 6

1. Letter to the author from Merritt A. Williamson, 7 August 1979.
2. "Annual Report of the College of Engineering and Architecture for 1956–57."
3. Interview with Arthur H. Waynick, State College, Pa., 8 May 1979.
4. Letter to the author from Merritt A. Williamson, 14 November 1979.
5. Floyd L. Carnahan, "Early History of Chemical Engineering at Penn State," 1976 (typewritten); Nunzio J. Palladino, "Merrill R. Fenske, 1904–71," in *Memorial Tributes* (Washington: National Academy of Engineering, 1979), pp. 50–56.
6. Venuto, "Steidle," pp. 86–87; Mary S. Neilly, "A Brief History of the College of Earth and Mineral Sciences," *Earth and Mineral Sciences* 50 (September 1980): 1–8.
7. "Annual Report of the College of Engineering and Architecture for 1959–60."
8. Interview with Nunzio J. Palladino, State College, Pa., 9 August 1979.
9. *Final Report: Goals of Engineering Education* (Washington: American Society for Engineering Education, 1968), pp. 9–10.
10. *Penn State Engineer,* May 1955, pp. 9, 20.

11. *Thirty-fifth Annual Report of the Engineers' Council for Professional Development* (New York: ECPD, 1967), p. 85; *Forty-sixth Annual Report of the Engineers' Council for Professional Development* (New York: ECPD, 1978), p. 14.

12. Interview with Nunzio J. Palladino, State College, Pa., 9 August 1979.

13. Typical cases are Philip S. Klein and Ari Hoogenboom, *A History of Pennsylvania,* 2nd ed. (University Park: The Pennsylvania State University Press, 1980); Wayland F. Dunaway, *A History of Pennsylvania,* 2nd ed. (Englewood Cliffs: Prentice-Hall, 1948); and Sylvester K. Stevens, *Pennsylvania: Birthplace of a Nation* (New York: Random House, 1964). Stevens' work is not a textbook in the strict sense. The official history of Penn State, Wayland F. Dunaway's *History of the Pennsylvania State College* (Lancaster, Pa.: n.p., 1946), gives agriculture and even science more emphasis than engineering in the University's development.

Appendix A

Deans of the College of Engineering

DEANS

Louis E. Reber, 1895–1907
John Price Jackson, 1907–1914
Elton D. Walker, 1914–1915 (acting)
Robert L. Sackett, 1915–1937
Harry P. Hammond, 1937–1951
Eric A. Walker, 1951–1956
Earl B. Stavely, 1956 (acting)
Merritt A. Williamson, 1956–1966
Howard L. Hartman, 1966 (acting)
Nunzio J. Palladino, 1966–1981
Wilbur L. Meier, 1981–

ASSOCIATE AND ASSISTANT DEANS

Royal M. Gerhardt, 1942–1948
Earl B. Stavely, 1948–1959
Kenneth L. Holderman, 1955–1959
Paul Ebaugh, 1957–1979
Lawrence J. Perez, 1957–1970
Francis T. Hall, Jr., 1959–1965
Albert H. Jacobson, Jr., 1959–1961
Arthur T. Thompson, 1961–1963
Howard L. Hartman, 1963–1967
Otis E. Lancaster, 1967–1974
Robert E. McCord, 1967–
Ernest R. Weidhaas, 1967–

Walter G. Braun, 1970–80
William H. Gotolski, 1974–
Winfred M. Phillips, 1979–80
George J. McMurtry, 1980–
Edward H. Klevans, 1980–

Appendix B

Engineering Department Heads

ARCHITECTURAL ENGINEERING / ARCHITECTURE

Roy I. Webber, 1911–19
Alfred L. Kocher, 1919–26
Clinton L. Harris, 1926–29 (acting), 1930–34
Louis F. Pilcher, 1929–30
Joseph M. Judge, 1934–35 (acting)
J. Burn Helme, 1935–36 (acting)
B. Kenneth Johnstone, 1936–46
Milton S. Osborne, 1946–62
Philip F. Hallock, 1962 (acting)
Gifford H. Albright, 1963–

Founded as the Department of Architectural Engineering, it became the Department of Architecture in 1922 with the introduction of a curriculum in architectural design. The architectural curriculum was transferred to the College of Arts and Architecture in 1963. Architectural Engineering then became a separate department within the College of Engineering.

AEROSPACE ENGINEERING

David J. Peery, 1944–54
Harold M. Hipsh, 1954–57
Irving Michelson, 1957–60
John A. Fox, 1960–61 (acting)
George F. Wislicenus, 1961–69
Barnes W. McCormick, 1969–

Formerly the Department of Aeronautical Engineering, its name was changed to Aerospace Engineering in 1966.

AGRICULTURAL ENGINEERING

Ralph U. Blasingame, 1920–50
Arthur W. Clyde, 1951–54 (acting)
Frank W. Peikert, 1954–75
Howard D. Bartlett, 1975–76 (acting)
Harold V. Walton, 1976–

Founded as the Department of Farm Machinery, its name was changed in 1930 to Agricultural Engineering. It did not become part of the College of Engineering until 1954.

CHEMICAL ENGINEERING

Donald S. Cryder, 1948–57
Floyd L. Carnahan, 1957–58 (acting)
W. C. Fernelius, 1958–59 (acting)
Merrell R. Fenske, 1959–69
Walter G. Braun, 1969–70 (acting)
Lee C. Eagleton, 1970–

This Department from 1904 until 1948 was a curriculum of the Department of Chemistry. It did not join the College of Engineering until 1963.

CIVIL ENGINEERING

Louis H. Barnard, 1881–93
Fred E. Foss, 1893–1907
Elton D. Walker, 1907–39
Frederic T. Mavis, 1939–44
Raymond O'Donnell, 1944–46 (acting)
Benjamin A. Whisler, 1946–72
Raymond E. Untrauer, 1972–1979
Robert M. Barnoff, 1979–

ELECTRICAL ENGINEERING

John Price Jackson, 1893–1908
Charles E. Kinsloe, 1908–44
Earl B. Stavely, 1944–45 (acting)
Eric A. Walker, 1945–51
Arthur H. Waynick, 1951–71
William J. Ross, 1971–79
Dale M. Grimes, 1979–

ENGINEERING SCIENCE

John W. Breneman, 1953–55
Warren E. Wilson, 1955–57
John R. Mentzer, 1957–74

This Department was consolidated with the Department of Engineering Mechanics in 1974 to form the Department of Engineering Science and Mechanics.

ENGINEERING SCIENCE AND MECHANICS

Charles E. Paul, 1907–08
Paul B. Breneman, 1908–38
Rudolph K. Bernhard, 1938–44
Raymond O'Donnell, 1944–46 (acting)
John A. Sauer, 1946–53
Joseph Marin, 1953–65
Robert K. Vierck, 1965–67 (acting)
Robert M. Haythornethwaite, 1967–74
John R. Mentzer, 1974–81
Richard P. McNitt, 1981–

Founded as the Department of Mechanics and Materials of Construction, it became the Department of Engineering Mechanics in 1938. In 1974 it was consolidated with the Department of Engineering Science.

GENERAL ENGINEERING

Francis T. Hall, Jr., 1959–65
Ernest R. Weidhaas, 1965–

This Department encompasses undergraduates (mainly freshmen) at the University Park campus who have not declared a major, and all faculty and students at the Commonwealth Campuses, including those in the associate degree programs.

INDUSTRIAL AND MANAGEMENT SYSTEMS ENGINEERING

Hugo Diemer, 1909–19
Edward J. Kunze, 1919–21
J. Orvis Keller, 1921–25
Charles W. Beese, 1925–30
Clarence E. Bullinger, 1930–55
Benjamin W. Niebel, 1955–78
William E. Biles, 1978–81
Allen L. Soyster, 1981–

Founded as the Department of Industrial Engineering, it assumed its present name in 1973.

MECHANICAL ENGINEERING

Louis E. Reber, 1886–1907
Hugo Diemer, 1907–09
Louis A. Harding, 1909–12
James A. Moyer, 1912–15
Edward P. Fessenden, 1915–21
Arthur J. Wood, 1921–31

Harold A. Everett, 1931–46
Norman R. Sparks, 1946–60
Maurice A. Gjesdahl, 1960–62
Richard G. Cunningham, 1962–71
Donald R. Olson, 1971–

NUCLEAR ENGINEERING

Nunzio J. Palladino, 1959–66
Anthony H. Foderaro, 1966–67 (acting)
Donald H. Kline, 1967 (acting)
Warren F. Witzig, 1967–

ENGINEERING EXPERIMENT DEPARTMENT

Arthur J. Wood, 1918–21
F. G. Hechler, 1921–50
Maurice Nelles, 1950–51
Elmer R. Queer, 1951–64

Originally the Engineering Experiment Station, it became the Department of Engineering Research in 1951 and the Engineering Experiment Department in 1957. From the Station's founding in 1910 until 1918, the Dean served as official director while the head of the Department of Mechanical Engineering was in immediate charge.

CONTINUING EDUCATION IN ENGINEERING

James A. Moyer, 1912–15
Norman C. Miller, 1915–25
J. Orvis Keller, 1925–34
Edward L. Keller, 1934–49
Kenneth L. Holderman, 1949–59
Robert E. McCord, 1959–64

Founded as the Engineering Extension Department, it became the Department of Continuing Education in Engineering in 1959 and was terminated as an independent department in 1964.

Appendix C

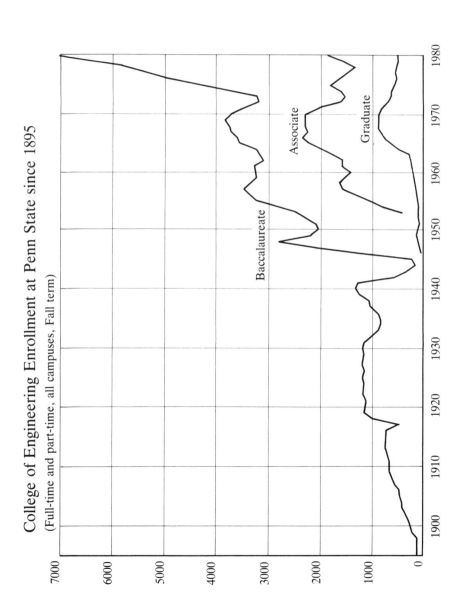

Appendix D

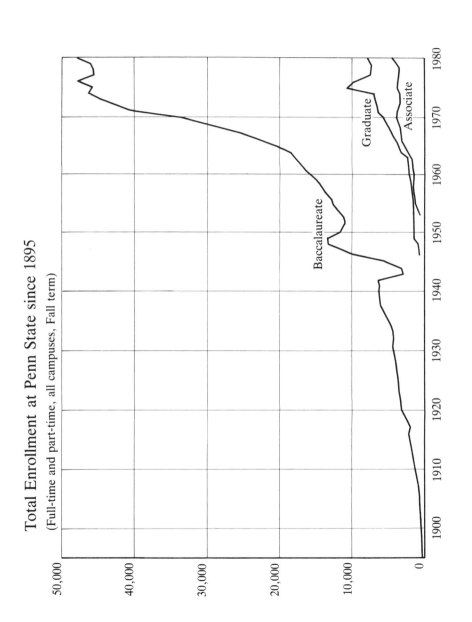

Total Enrollment at Penn State since 1895
(Full-time and part-time, all campuses, Fall term)

Index

Accreditation. *See* Engineers' Council for Professional Development
Aeronautical engineering curriculum, 136–38, 144–45, 151, 153, 183, 204
Aerospace curriculum, 204
Agricultural College of Pennsylvania. *See* Pennsylvania State University, The
Agricultural engineering curriculum, 176–79, 183, 204, 210
Agriculture, College (School) of, 33, 116, 124, 176–78
Albright, Gifford, 192–93
Alexander, C. T., 13
Allegheny College, 5
Allen, William H., 5–6
Allentown branch school, 79, 92
Altenhof and Brown, 159
American Chemical Society, 72
American Society for Engineering Education, 32, 160, 171–72, 175, 189, 202–03. *See also* Society for the Promotion of Engineering Education
American Society of Civil Engineers, 65, 72, 136
American Society of Heating and Ventilating Engineers, 93
American Society of Mechanical Engineers, 65, 72, 112, 121–22
American Society of Refrigerating Engineers, 112–13
Applied Research Laboratory, 199, 202. *See also* Ordnance Research Laboratory
Architectural engineering curriculum, 57, 59, 95, 100, 104, 116, 134, 155, 183, 192–94, 204, 210
Architecture curriculum, 99–100, 108, 110, 116, 129, 134, 141, 153, 155, 159, 175, 183, 192–93
Argonne National Laboratory, 168
Arts and Architecture, College of, 192–93
Associate degree programs, 170–72, 178, 206–07, 210
Association of American Agricultural Colleges and Experiment Stations, 64–65
Astronomy, Department of, 188

Index

Atherton, George W., 13–14, 16–17, 20–24, 26, 32–34, 40, 42–44, 52–53
Atomic Energy Commission, 165–68, 184, 207

Barnard, Louis, 12, 14, 24, 29, 33
Bates, Edward, 82–83
Beaver, James A., 23, 57, 59
Beese, Charles W., 106, 113, 123, 171
Bellefonte Central Railroad, 38, 46–47, 61, 66, 74–75, 150
Bernhard, Rudolph K., 135, 140–41
Beta Rho Delta, 203
Beyer, T. Raymond, 29
Blasingame, Ralph U., 176–77
Boerlin, Irving C., 123
Branch campuses. See Commonwealth Campus system
Breazeale, William M., 165–68, 185
Breneman, John W., 173
Breneman, Paul B., 56, 71, 113, 135
Brooklyn Polytechnic Institute, 128
Brumbaugh, Martin W., 70, 81
Bullinger, Clarence, 113, 123, 139
Burrowes, Thomas H., 7–8
Bush, E. W., 71
Butts, Edward P., 31, 71

Calder, James, 9–11, 13
California, University of, 41
Campbell, J. M., 11
Capitol Campus, 206
Carnegie, Andrew, 64
Carnegie Institute of Technology, 49
Chemical Engineering Building, 190, 196
Chemical engineering curriculum, 194–98, 204
Chemistry and Physics, College (School) of, 118–19, 122, 124, 129, 135, 194
Chesney, Cummings C., 30–31
Churchill, Jesse B., 194
Cincinnati, University of, 136
Civil engineering curriculum, 3, 6–9, 12, 14–16, 21, 26, 29, 38, 43, 47, 56, 71, 86, 88, 95, 102–03, 108, 110, 112, 116, 131–32, 135, 153, 183, 204
Climatometer, 144
Clyde, Arthur W., 177
Columbia University, 44, 155
Committee on Special Education and Training, 82
Commonwealth Campus system, 125, 151, 158, 184, 191, 210–11
Commonwealth Industrial Research Corporation, 199
Computer Science, Department of, 187
Computers. See PENNSTAC

Continuing Education in Engineering, Department of, 191, 210. *See also* Extension education
Cooperative degree program, 169–70. *See also* Dual degree program
Cooperative education, 136
Corliss steam engine, 24, 188–89
Cornell University, 41, 58, 122, 155
Correspondence instruction, 69, 91–92, 123–24
Cresswell, Donald, 101
Crossley, Gilbert H., 101
Cryder, Donald, 195
Curtiss–Wright Corporation, 142, 145, 157, 185

Darlington, Frederick, 31
Dartmouth College, 3, 155
Day and Klauder, 88, 109
Dedrick, Benjamin F., 63, 83, 104
DeJuhasz, Kahlman J., 111, 190
Denney, Charles E., 50, 160
Dickinson College, 5
Diemer, Hugo, 57–58, 75, 82, 90
Diesel engineering, 105, 111–12, 118, 129, 139, 143, 153, 190
Doherty, Robert E., 147
Downey, John F., 9, 11
Drawing and Descriptive Geometry, Department of, 56
Driver education, 120–21
Dual degree program, 170. *See also* Cooperative degree program
Duff, James H., 151

Earle, George H., 129–30
Ebaugh, Paul, 184
Education, College (School) of, 100, 106, 124, 175
Eisenhower, Dwight D., 161, 166–68
Eisenhower, Milton S., 161–62, 165–67, 169, 179, 181, 198
Electric railway, experimental, 38–39, 45–46
Electrical Engineering Building East, 190
Electrical Engineering Building West, 130
Electrical engineering curriculum, 20–21, 25–29, 35, 38–39, 42–46, 60–63, 71, 87, 95, 101, 104, 116, 129–30, 142, 153–54, 183, 187–88, 204
Electrochemical engineering curriculum, 45, 95–96, 104, 116, 134
Elliot Fellowship, 121
Ellsworth, Clarence, 167
Emergency Building Fund Campaign, 96, 98–99
Engineer, The, 73–74, 93–94
Engineering, College (School) of: admission requirements, 49, 80–81, 89–91; alumni, 30–32, 50–51, 71, 95, 103–04, 160, 192, 197; compared with other institutions, 41, 44, 56, 90–91, 104, 153, 155, 197, 206; general curricular development, 29–30, 32–

232   *Index*

37, 42, 51–52, 72–73, 126, 132–35, 155, 174–75, 204; name changed, 135, 169, 194; student enrollment, 20–21, 26, 37–38, 40, 43, 48–49, 110, 115–16, 138, 143, 147, 150–51, 156, 183, 201–04; workload and salaries of faculty, 38, 43, 78, 90, 98, 109, 141, 152, 164–65
Engineering College Magazines Association, 117
Engineering College Research Association, 160
Engineering Drawing, Department of, 56, 59
Engineering Experiment Department, 184, 193
Engineering Experiment Station, 65–66, 78, 83, 93, 97, 104–05, 108, 112, 122, 129, 136, 143–44, 153, 156, 163–64, 194, 211
Engineering Mechanics, Department of, 135, 174, 183
Engineering Mechanics and Materials, Department of, 56, 78, 135
Engineering Research, Department of, 164, 184
Engineering, Science, and Management Defense Training Program, 139–41, 145
Engineering Science and Mechanics, Department of, 174
Engineering science curriculum, 172–74, 183, 204
Engineering technology curriculum, 206
Engineering Units: A, 87–88; B, 81, 84, 87; C, 87; D, 61, 78, 87, 95, 129; E, 60–61, 78, 87, 95–96, 129–30, 144; F, 59–61, 63, 78, 86–87, 129, 159
Engineers' Council for Professional Development, 125–28, 133, 171, 177–78
Engineers Joint Council, 206
Environmental engineering curriculum, 204, 206. *See also* Sanitary engineering curriculum
Everett, Harold A., 118–19, 123, 136–37, 139
Extension education, 66–69, 78–80, 92, 97, 105–06, 112, 123–25, 139, 157–58, 162–63, 170–72, 210. *See also* Continuing Education in Engineering, Department of

Farm Machinery, Department of, 176–77
Farmers' High School. *See* Pennsylvania State University, The
Fenske, Merrell R., 195–97
Fenske Laboratory. *See* Chemical Engineering Building
Fessenden, Edward P., 80, 88, 97
Fine arts curriculum, 99, 175
Fisher, John S., 109, 111
Foss, Fred E., 25, 29, 44, 75
Fraser, John, 6–8

Garfield Thomas Water Tunnel, 154–55
General Engineering, Department of, 191, 204
General State Authority, 129–30, 137–38, 159, 185, 189, 193
George Westinghouse Professorship of Engineering Education, 173–74, 188–89
George Westinghouse Professorship of Production Engineering, 173
Gerhardt, Royal, 141, 145, 152
Gildea, David J., 144
Gill, Arthur H., 67
Gillan, Gerald K., 167

Index   233

Good Roads Train, 69-70
Gossler, Philip G., 31
Govier, Charles, 113
Graduate studies, 71-72, 112, 121-22, 142, 155-56, 158, 162-63, 172-73, 203-05
Grinter Report, 175, 179
Guyer, Arthur G., 29

Hagen, John P., 188
Hall, Francis T., Jr., 191
Hamill, James L., 54
Hamilton Standard Division of United Aircraft, 142, 157
Hammond, Harry P., 127-40, 143, 145, 147-52, 155, 157-58, 160-61, 165, 172, 174, 177, 179, 190
Hammond Building, 17, 42, 189-90
Hammond Report: of 1940, 134, 148, 175; of 1944, 147-49, 170-71, 175
Harding, Charlotte Hanes, 75
Harding, Louis A., 66, 74-75
Harris, Clinton L., 113, 120
Hartman, Howard, 199
Harvard University, 3, 149
Healy, John F., 16
Hechler, Frederick George, 97, 104-05, 113
Heilman, Israel, 8
Hench, John B., 29
Hetzel, Ralph D., 109, 119-20, 124-25, 128-30, 137, 145, 150, 152
Hetzel, Theodore B., 122
Highway engineering curriculum, 63
Hoffman, Samuel K., 144
Holderman, Kenneth, 158, 170-72, 191
Honors program, 174
Houston, University of, 171
Howell Lewis Shay and Associates, 189-90
Hunter, Esther, 10
Hunter and Caldwell, 130
Hussman, A. W., 184

Ihlseng, Magnus C., 25, 40, 47
Illinois, University of, 41, 65, 121
Industrial and Fine Arts, Department of, 99
Industrial and Management Systems Engineering, Department of, 206
Industrial and Professional Advisory Council, 191-92
Industrial chemistry. *See* Chemical engineering curriculum
Industrial Design, Department of, 99
Industrial education curriculum, 58-59, 100
Industrial engineering curriculum, 57-58, 82, 87-88, 90, 98, 112, 116-17, 136, 139, 151, 183, 204, 210

234 *Index*

Inspection trips, 102, 117
Institute for Building Research, 193
Institutional Advisory Service, 184
Institutional Engineering Service, 193–94
International School of Nuclear Science and Engineering, 168
Ionosphere Research Laboratory, 153–54, 161, 163, 175, 188, 211
Iowa State College, 41, 65, 177

Jackson, Dugald C., 31–32
Jackson, John Price, 20, 25–26, 29, 31, 38, 41, 44, 50, 54–56, 61–62, 65, 67–73, 75, 77, 80, 174, 208
Jackson, Josiah, 11–12, 26
Jackson, Lyman E., 177
Jordan, Whitman H., 17

Kapp, Paul B., 93
Kaulfuss, J. E., 120
Keller, Edward L., 125, 158, 191
Keller, John Orvis, 82, 97, 106, 113, 123–25, 158
Kennedy, Alfred, 5
Killmer, Miles I., 51
Kimball, Dexter, 58
Kinkaid, Thomas W., 27
Kinsloe, Charles L., 48, 66, 75, 82, 113, 141, 149
Kirby, Robert E., 197
Kocher, Alfred L., 90, 104, 109
Koehler, John T., 155
Kountz, Rupert, 167
Kunkle, Bayard D., 50–51, 160, 189–90
Kunze, Edward J., 90, 97

Lafayette College, 5, 49, 56
Lamme Award, 126, 160, 179
Lancaster, Otis E., 188–89
Land-Grant Act. *See* Morrill Land-Grant Act
Land-Grant College Engineering Association, 65
Lardner, Henry A., 29
Leader, George, 172, 190
Lehigh University, 10, 49, 56
Liberal Arts, College (School) of, 99, 106, 116, 124, 135, 170, 175, 198
Loomis, William P., 100
Lutz, Sherman, 136–37

McAllister, Addams S., 50
McCord, Robert E., 191
McCormick, John H., 12

McElwain, Carrie, 31, 142
McElwain, Harriet, 31
McGraw Award, 172
McKee, Arthur G., 31
McKee, James Y., 12
MacKenzie, Ossian R., 187
Main Engineering Building, new. *See* Sackett Building
Main Engineering Building, old, 21–26, 38, 40, 42, 52, 64, 73, 78, 84–86, 89, 94
Maine, University of, 177
Mann Report, 72, 132
Manson, Lidia, 142–43
Massachusetts Institute of Technology, 16–17, 23, 32, 41, 44, 122
Mavis, Frederic T., 135–36, 141
Maxwell, Charles E., 71
Mechanic Arts Building, 17–18, 48, 129
Mechanic arts curriculum, 12, 16–18, 21, 27, 43, 58
Mechanical engineering curriculum, 6–7, 17–21, 26, 39, 42–43, 46–47, 56, 62–63, 71, 86, 88, 95, 117, 136, 138, 142, 144, 151, 153, 183, 204
Mechanical Engineering Laboratory (Building), 87–89, 105, 129–30, 136, 144, 151–52, 159, 190
Meier, Wilbur L., 208
Mentzer, John R., 174
Meyer, Wolfgang E., 184
Michigan, University of, 3, 44
Michigan State University, 177
Miles, H. E., 20
Milholland, James, 151, 159
Miller, Norman C., 80, 91–92, 106
Milling engineering curriculum, 63, 83, 104, 116–17, 210
Mineral Industries, College (School) of, 118–19, 124, 130–31, 198–99
Mines, School of, 35, 40–41, 47, 78, 100, 106, 119, 198
Mining engineering curriculum, 5–7, 25–26, 40–41, 46–47, 198
Minnesota, University of, 17
Missouri, University of, 41
Mock, James C., 31
Moore, Elwood S., 100
Morrill Land-Grant Act: of 1862, 3–11, 13, 21, 32, 90, 209; of 1892, 32
Moyer, James A., 68–69, 80, 106

National Academy of Engineering, 179
National Aeronautics and Space Administration, 184
National Architectural Accrediting Board, 155
National Council of State Boards of Engineering Examiners, 125
National Science Foundation, 160, 163, 184, 186, 188
Natural Science, School of, 194
Naval architecture and marine engineering curriculum, 137

Needham, Anne, 197
Neyhart, Amos E., 120–21
Nicholas, John E., 177
Noble, John W., 23
Norris, Earl B., 151
North Carolina State College, 166–68
Norwich University, 3
Nuclear engineering curriculum, 165, 168–69, 185, 200–01, 204, 210
Nuclear reactor facility, 165–68, 185–86
Nuclear Regulatory Commission, 207–08

Oak Ridge National Laboratory, 165, 167
Office of Scientific Research and Development, 149–50
Ohio State University, 41
Old Main, 9, 16, 18, 23, 64
Olds, Fred L., 23
Ordnance Research Laboratory, 150–51, 153–55, 159, 161, 163, 199, 211. *See also* Applied Research Laboratory
Orvis, Ellis, 54
Osborne, Milton, 152
Osmond, I. Thornton, 11–12, 20, 25
Oswald, John W., 204

Palladino, Nunzio J., 185, 200–01, 207–08
Pattison, Robert E., 23–24, 53
Paul, Charles E., 56
Peery, David J., 138, 144–45, 174
Peikert, Frank W., 177–78
Pemberton, John, 27
Penn State Engineer, 93–94, 101–02, 110, 117–18, 133, 143, 156–57, 168, 192
Penn State Engineering Society, 192
PENNSTAC, 186–88
Pennsylvania, University of, 41, 49, 56, 107, 186
Pennsylvania Department of Highways, 66, 69, 120
Pennsylvania Grade Crude Oil Association, 119, 195–96
Pennsylvania Railroad, 25, 38, 46–47, 50, 61–62, 67–69, 86, 158
Pennsylvania State Agricultural Society, 1
Pennsylvania State Chamber of Commerce, 89–90
Pennsylvania State College. *See* Pennsylvania State University, The
Pennsylvania State Millers Association, 63, 83
Pennsylvania State University, The: established, 1–2; name changed, 4, 9, 169; financial and administrative relationship with College of Engineering, 32–35, 37–38, 40, 53, 60–61, 80–81, 89, 94–100, 107–09, 111, 119–20, 129–30, 150–52, 160–61, 165–66, 179, 192–94, 198, 209–11
Pennsylvania State University Engineering Association, 192
Pennypacker, Samuel W., 54

Index   237

Peters, Max S., 197
Petroleum Refining Laboratory, 129, 190, 195–97
Philadelphia Traction Company, 38
Physics and Electrotechnics, Department of, 20–21, 25–26
Pilcher, Louis, 120
Pinchot, Gifford, 98
Pittsburgh, University of, 107. *See also* Western University of Pennsylvania
Polytechnic College of the State of Pennsylvania, 4–5
Public Works Administration, 120, 129–30
Pugh, Evan, 2–5
Purdue University, 41, 77, 121, 171, 177
Pyle, Hugh G., 123

Quarles, Gilford G., 161
Queer, Elmer, 164, 193
Quiggle, Dorothy, 197

Radio engineering, 62–63, 100–01, 121
Railway mechanical engineering curriculum, 39, 46–47, 61–62, 104, 116–17
Reber, Louis E., 16–18, 21–25, 27, 30, 34–35, 39–44, 46, 51–54, 56–57, 70, 105–06, 174, 208
Redifer, Anna E., 99
Rennselaer Polytechnic Institute, 3, 97, 122
Research in engineering, 64, 97, 108–09, 118–21, 149, 153, 163–65, 183–85, 205, 211.
    See also Diesel engineering; Engineering Experiment Station; Thermal laboratories
Robinson, Francis, 8–9
Rockview State Penitentiary, 106

Sackett, Robert L., 77–82, 84, 88, 90, 93, 95–97, 99–100, 104, 107–09, 111, 113, 115, 117, 119–20, 122–23, 126–29, 131, 136, 174, 190, 198
Sackett Building (New Main Engineering Building), 109–10, 128–29, 151–52, 159, 189
Sanitary engineering curriculum, 29, 47, 86, 116, 183, 206, 210. *See also* Environmental engineering curriculum
Schilling, Harold K., 163
Schneider, Herman, 136
Schwab, Charles M., 42–43
Schweitzer, Paul, 105, 111, 118, 122, 142, 153, 190
Science, College of, 187–88, 194
Shaad, George C., 51
Shattuck, Harold, 82
Shields, John Franklin, 31
Shortlidge, Joseph, 11–12
Slater, William L., 197
Smith, Olga, 142
Society for the Promotion of Engineering Education, 32, 55, 72, 102, 125–26, 128, 132, 134, 147, 160. *See also* American Society for Engineering Education

Society of Women Engineers, 203
Sparks, Edwin E., 54–56, 59–60, 63, 65, 70, 72, 77, 81, 83–85, 89
Sparks, Norman R., 164
Spectrum, 118
Sproul, William, 94
Starkweather, S. W., 13
Stavely, Earl B., 152–53
Steidel, Edward, 198
Stevens Institute of Technology, 17
Stoll, Clarence G., 50, 160
Stone Valley summer camp, 131–32
Strauss, Lewis L., 167
Struble, Jacob, 20, 31
Student Army Training Corps, 83–84
Student engineering societies, 73, 101–02, 126, 203
Student publications. See *The Engineer, Penn State Engineer, Spectrum*
Summer camp, 102–03, 131–32
Summer Institute for Effective Teaching for Engineering Teachers, 188

Tarpley, Harold I., 187
Taylor, Frederick, 57
Technical institutes, 157, 170–72, 175, 210
Temple University, 107
Tener, John K., 60, 70, 75, 109
Texaco Fellowship, 121–22
Thermal laboratories, 66, 78, 136, 184
Thomas, John M., 94–96, 98–100, 106–08
Thornburgh, Dick, 207
Towle, William M., 27

Undergraduate centers. *See* Commonwealth Campus system
Underwater Sound Laboratory, 149–50
United States Bureau of Education, 107
United States Bureau of Roads, 69
United States Department of Agriculture, 63, 131
United States Navy, 27, 112, 139–40, 143–44, 149–50, 154, 159
Urey, Harold, 195

Van Hise, Charles, 53
Very, Ann, 142
Very, Dexter, 142

Wadsworth, Marshman E., 44, 47–48
Walker, Elton D., 47, 66, 75, 77, 82, 93, 113, 131, 135
Walker, Eric A., 149–50, 160–75, 179–81, 189–92, 199, 203–04
Walker, Francis A., 23

Walker, William H., 194–95
Washington University, 17
Watts, Ralph L., 177
Waynick, Arthur H., 154, 161
Webber, Roy, 90
Weidhaas, Ernest R., 203
Western University of Pennsylvania, 5, 7, 10, 41, 49, 56. *See also* Pittsburgh, University of
Whitmore, Frank C., 195
Whitmore, James B., 56
Wickenden Report, 128, 132–34
Wickersham, J. P., 11–12
Williamson, Merritt A., 181–84, 188–93, 198–201
Williamsport Vocational School, 67–68
Wilson, Warren E., 173–74, 188
Wisconsin, University of, 32, 41, 51, 53, 70
Witman, Clyde, 167
Wood, Arthur J., 47, 62, 73, 78, 93, 97, 104–05, 112–13, 123
Woodruff, Eugene C., 100–01, 121
Worcester Polytechnic Institute, 17, 23
WPAB, 101
WPSC, 101

Yale University, 3
York Ice Machinery Company, 136